Rock, Gem, and Mineral Collecting Sites In Western North Carolina

53 of the best sites in the area

Maps and GPS Coordinates!

Restrictions, Owners, Fees

What you need, what you get

Rick James Jacquot, Jr.

Foreword by Ralph Roberts

Land of the Sky Books

Alexander, North Carolina

Publisher: Ralph Roberts
Editor: Pat Roberts
Cover Design: Ralph Roberts
Interior Design & Electronic Page Assembly: **WorldComm**®

10 9 8 7 6 5 4 3 2 1

ISBN 1-56664-247-7 trade paper
ISBN 1-56664-248-5 hardback

Land of the Sky Books™—a division of Creativity, Inc.—is a full–service publisher located at 65 Macedonia Road, Alexander NC 28701. Phone (828) 252–9515, Fax (828) 255–8719. For orders only: 1-800-472-0438. Visa and MasterCard accepted.

Land of the Sky Books™ is distributed to the trade by **aBOOKS™** 65 Macedonia Road, Alexander NC 28701. Phone (828) 252–9515, Fax (828) 255–8719. For orders only: 1-800-472-0438.

This book is also available on the Internet at **abooks.com**
Set your browser to **http://abooks.com** and enjoy the many fine values available there.

CONTENTS

*"Many people take the mountains for granite,
but they are really gneiss."*
—very old geology joke

Foreword

There are things you can collect in these old, old mountains that have greater age than antiques. Far greater age. Eons old, and these treasures lie beneath your feet, in the ground, the ancient ground.

Rick Jacquot has knowledge of these vastly old artifacts of the earth's creation and its growing pains through millions of long years, now gone. He has searched over mountain and through bramble-choked glen to find the best places, those hallowed locations yielding the best in specimens of rock, mineral, and—oh yes!—sparkling gemstones. Some of these specimens can be valuable, others precious in the learning of geological lore they impart. All have a story to tell.

Often hunters of rocks maintain their secrets as closely as any fisherman protecting the piece of stream where the big trout grab for any hook that comes near the water. Rick does not, he shares it here with you, even to giving GPS coordinates!

When Rick brought this book idea to me, I turned out to be an easier sell than he had thought. My father and I spent many happy hours in the pursuit of the not-always-elusive rock. My cousins—Jack Ball and his son Jackie—have ownership of the Little Pine garnet mine in Madison County where my grandfather, George Roberts, was foreman back in its heyday before World War II. I love these mountains—what is on them, in them, and what makes them up. So we, in much pride, add this book to our **Land of the Sky** series.

—Ralph Roberts, Publisher

Acknowledgements

I would like to thank the following people for helping with the history and site information in this book.

My son R.J. Jacquot who helped me gather information.

Bill Mintz, friend and fellow rockhound who provided information on to many sites to list.

Steve Penley, friend and fellow rockhound who provided information for the Carter Corundum Mine, Black Mountain Kyanite, and the Erwin Tennessee Calcite site.

Jack Ball Sr., owner of the Little Pine Garnet Mine.

Jack Herbert, president of the Nantahala Talc and Limestone Company.

Ed Silver, owner of the Sinkhole Mine.

Terry Ledford, owner of the Crabtree Emerald Mine.

Mitchell County Chamber of Commerce.

Branson Woods, owner of the Woods Creek Sapphire Mine.

Bruce Caminiti, for information on the Woods Creek Mine.

Effie McCrackine, owner of the Cherokee Ruby Mine.

Brown Johnson, owner of the Mason Mountain Mine.

Barry and Tonya Parker, for information on the Coles Crossing Georgia site.

Tiffany Poovey, owner of the Poovey Garnet site.

Books

Mines Miners and Minerals, Lowell Presnell.

Gems and Minerals of America.

Appalachian Mineral & Gem Trails, June Culp Zeitner.

Mineral Collecting Sites in North Carolina, W.F. Wilson and B.J. McKenzie.

Gemstones of North America Vol. III, John Sinkankas.

Dedication

For my son, R.J.
who shares the spirit of outdoor adventure
and exploration with me

Introduction

The Western North Carolina area has been mined/prospected off and on for a variety of gems and minerals as far back as the 16th century. I can only imagine what it must have been like to be one of those early prospectors, to be the first one to discover a gem bearing pegmatite or to find gem quality rubies and sapphires in the local creekbeds. Commercial/Systematic mining for various minerals began in the 1700s and in 1871 C.E. Jenks opened the first gem corundum mine

Over the years, improved mining techniques uncovered many more rich gem and mineral deposits. Unfortunately gem production was too low to justify continued commercial mining, mineral mines began to close as imported minerals began to be shipped into the area, it was cheaper to import the minerals from a foreign country than to mine them locally.

Today the gem mines you find are tourist mines/attractions, these mines are enriched with foreign material to keep the customers happy, most are located in the Cowee Valley and the Spruce Pine area, there are a few mines which provide all native material and I have listed them in this book. There are still some active commercial mines/quarries in Western North Carolina, limestone, talc, feldspar, quartz, pyrophyllite and spodumene, are just a few of the minerals mined.

The Western North Carolina area is still a rockhound's dream, if you know where to look. I moved here in 1987 and began collecting in this area in 1989. One thing I have learned about rockhounds is that they are tight-lipped about collecting sites, they do not want anyone to know where their secret hole is at, and if by chance you do get some information from them, you have to swear on your life, and promise to forfeit your firstborn child if you ever tell anyone about the secret location. This can become very frustrating, fortunately I met a local rockhound who had been collecting in the area for over thirty years and he did not mind sharing his information. I was lucky, and as my interest in the hobby grew, I began to meet more people who knew about more sites. Unfortunately many people do not get that chance, I am sure many aspiring rockhounds have simply given up the hobby for lack of interest because they could not find anyone willing to share information with them.

There are thousands of mines and prospects in the Western North Carolina area containing many gems and minerals that would be of interest to collectors. Many of these mines are on state or federal property but the locations to them have been lost over time and are waiting to be rediscovered, many more are on private land and the owners do not allow collecting. Many of the active mines and quarries in the area will not let anyone in to collect mineral specimens, they site safety reasons, government rules and regulations, etc. This book was written by a rockhound for rockhounds new and old.

I have listed as many sites that I know of that are open to collecting, some of these sites are active mines/quarries, some are the remains of old pegmatite/mica/feldspar/gem mines that have been closed for years, and a few are more recent discoveries made by myself and some fellow rockhounds in the area. I have also included several close sites that are in Georgia, Tennessee, and farther east in North Carolina. I feel you should have the opportunity to collect at these sites while in the area. I have tried to make finding the sites as easy as possible with detailed directions and maps.

The odometer on my vehicle was used to the tenth of a mile to measure some of the distances in this book, your odometer may vary slightly from mine so keep this in mind while traveling to a particular site. Remember to always respect the property of others, if you dig a hole make sure to fill it before you leave and always carry out any trash you bring in and any trash you see left by others, doing this will help to insure future collecting at these locations. If you are new to the hobby I hope this guide will help you along the way to find many beautiful specimens for your collection, if you've been around a while I hope you will find some useful information here that is new to you.

Many of the sites listed in this book are commercial mines or quarries or roadside locations that are easily found with the street directions and maps I have provided. Most of the commercial mines and some of the quarries have signs directing you to the site when you are within a few miles of the location, but as many more are in remote hard to find locations. You should have no problem finding these sites with the directions given, as an extra tool, if you have a portable GPS unit I have included the coordinates for the more remote locations to assist with finding these sites.

While researching this book over a two-year period I visited all the locations listed. I try to visit these locations as often as possible to collect specimens for my collection and I have always been granted permission to collect at the sites on private property. But collecting status changes and property is bought and sold, you should always get permission from the property owner before collecting on private property, this shows respect for the owner and helps to allow future collecting at these sites.

—R.J. Jacquot

SAFETY

Safety while rock collecting should be your number one priority, I have tried to list any safety concerns I have about a collecting area or mine with each site description. If the area has railroad tracks, old mine shafts, high walls with falling rock etc. I would not bring any small children to that site, there are many places listed in this book where children will have fun and you will not have to worry.

When working in an old mine shaft, inspect your surroundings to see if it is stable, if you see water leaking through the roof of a mine shaft or the support timbers are rotted and collapsed, you may want to find another place to collect, remember that there is plenty of nice material outside the mine on the dump piles, where there is no danger of a cave in.

You should also be aware of old vertical mine shafts in the woods or near dump piles, some of these shafts are overgrown with trees and brush and can be hard to spot until you are right on top of them, some of these shafts are very deep, some are full of water so be careful. Always tell someone where you are going when visiting an old mine or collecting site, give them directions to the mine and tell them when you plan to return, if you have a cell phone, bring it with you.

Some of the locations I have listed in this book have a variety of wildlife around the collecting sites such as: bears, mountain lions, snakes, wild boars, etc. I have encountered these animals on several occasions while rock collecting in the mountains and have never had a problem with them, remember they are more afraid of you than you are of them, and remember, never try to pet a mountain lion or wild hog or try to wrestle a bear—they usually win. There are times when animals may become aggressive. I prefer a .357 or .45 cal. animal repellant just in case one of these critters thinks I look like his lunch. You should also keep a first aid kit in your vehicle with a snakebite kit inside. Some of the locations listed in this book are a long way from any hospital. I am not telling you these things to discourage you from visiting these sites; I think you should visit any remote collecting area armed with as much information about that area as possible so there won't be any surprises.

Happy hunting.

A LIST OF THINGS TO BRING

Rock chipping hammer
3-lb. sledgehammer
6-lb. sledgehammer
15-lb. sledgehammer
Rock pick
Rock chisels, flat and round
Pry bar, short and long (4-6 feet)
Flat screwdriver
Shovel
5-gal. bucket
1/8" sifting screen
1/4" sifting screen
1/2" sifting screen
Specimen containers
Newspaper to wrap specimens
Bug spray (Off, Cutter, etc.)

Hard hat
Headlamp, lantern
Work gloves
Heavy boots
Safety glasses
Long pants
Rubber boots
Waders
Change of clothes
Camera
Compass
Maps
First aid kit with snakebite kit
Drinking water
Food

This book—so you will know where to go.

ROCKHOUNDING ON NATIONAL FOREST LAND IN NORTH CAROLINA

A wide variety of igneous, sedimentary and metamorphic rock types are found within the national forests in North Carolina, and many individual minerals are found in association with these rocks. As a rule, there is no objection to taking a handful of rock, mineral, or petrified wood specimens from the surface of national forest lands. No fee, special permission, or permit is required as long as the specimens are for personal non-commercial use, and the specimens are not of archeological value (all artifacts, projectile points, chips and flakes may not be collected).

No mechanical equipment is permitted and blasting is not allowed, no significant land / surface disturbance can result from collecting, and collecting cannot conflict with existing mineral permits, leases, claims, or sales. Surface/land disturbance is considered significant when, natural recovery would not be expected to take place within a reasonable period of time, there is unacceptable air or water degradation, there is unnecessary or unreasonable injury, loss or damage to national forest resources—i.e.. use of explosives or mechanical equipment.

This information obtained from:

Office of National Forests in North Carolina
160-A Zillicoa St.
Asheville, North Carolina
28802

Phone: 828-257-4200

LIST OF MINERALS FOUND AT THE MINES IN THIS BOOK

There are 55 rock, gem, and mineral collecting sites in this book. This is a list of the primary minerals you will find at these locations.

Actinolite (large crystals) up to 4"
Actinoilte (small green crystals)
Amazonstone (pale green)
Anatase crystals (blue)
Apatite (massive, short wave fluorescent)
Apatite (pale green crystals in matrix)
Apatite (dark green crystals in matrix)
Aquamarine crystals
Autunite (yellow/green tabular crystals, short wave and long wave fluorescent)
Azurite (dark blue micro crystals on matrix)
Beryl (large crystals up to 12", blue, green, gold, white)
Calcite (dogtooth crystals)
Cassiterite (near Yates Brooks Farm)
Cerussite (clear and white crystals up to 1/2")
Columbite (Sinkhole Mine)
Common opal
Corundum (star, asterism)
Corundum (with sillimanite)
Corundum (various colors, crystals)
Diamond (Propst Mine)

Dolomite (Nantahala Quarry)
Emerald crystals
Emerald in matrix
Feldspar (white, pink, yellow)
Galena (Redmond Mine)
Garnet (12 & 24 sided specimens)
Garnet (almandine)
Garnet (pyrope)
Garnet (rhodolite)
Hematite (black, botryoidal)
Hiddenite (Spodumene)
Hornfel (Lake Chatuge)
Hyalite opal (green & blue, short wave fluorescent)
Iridescent minerals
Kimberlite (Propst Mine)
Kyanite (massive pale blue)
Kyanite (large dark blue crystals)
Kyanite (crystal masses with sapphire)
Limonite (pseudomorph cubes)
Malachite (dark green micro crystals on matrix)
Marble (white, pink, yellow, blue, purple)
Margarite (Pressley Sapphire Mine)
Mica (biotite)
Mica (muscovite)
Mica (schist)
Moonstone
Nickeline
Olivine
Phenakite (Herbert Mine)
Pyrite (cubes)
Pyromorphite (green)

Quartz (chalcedony, various colors)
Quartz (with rutile inclusions)
Quartz (crystals, clear, smoky, phantom)
Quartz (massive)
Ruby (cabbing and facet grade, long wave fluorescent)
Ruby (in smaragdite matrix)
Rutile (crystals)
Sapphire (cabbing and facet grade)
Serpentine (Carter Mine)
Spinel (black crystals)
Spinel (massive, variety pleonast)
Staurolite (single crystals and crosses)
Talc (soapstone)
Thulite (pink, fluorescent)
Torbernite (dark green tabular crystals)
Tourmaline (black, schorl)
Tourmaline (green crystals)
Tremolite (asbestos) (Lake Chatuge, Grimshawe Mine)
Vermiculite
Zircon (fluorescent)
Zoisite

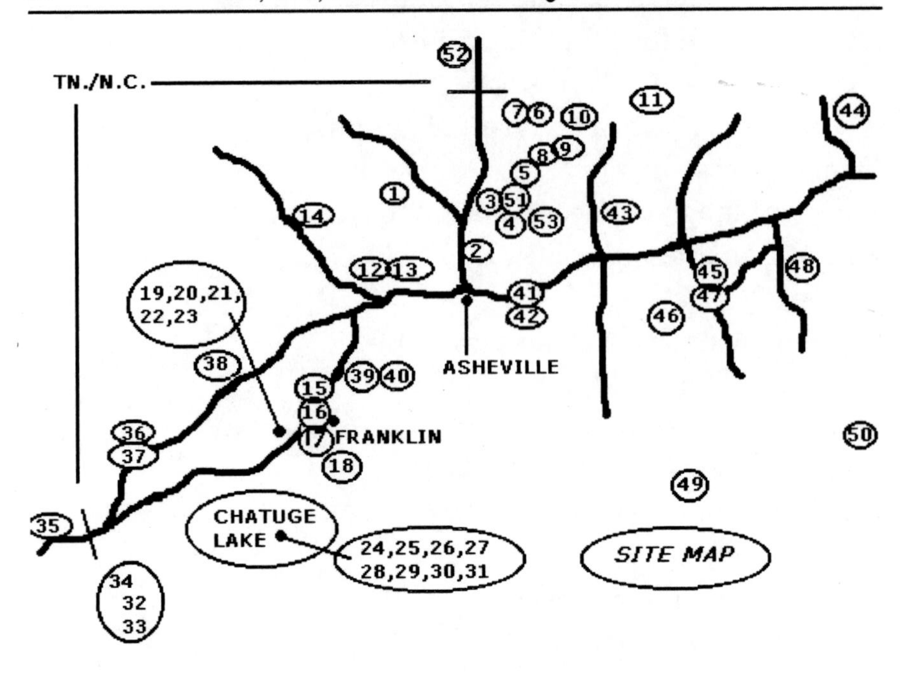
TN./N.C.
52
7 6
10
11
44
8 9
5
1
3 51
4 53
43
14
2
12 13
45
48
19,20,21,
22,23
41
47
42
46
38
ASHEVILLE
15
39 40
16
36
FRANKLIN
50
37
17
18
49
35
CHATUGE
LAKE
24,25,26,27
28,29,30,31
SITE MAP
34
32
33

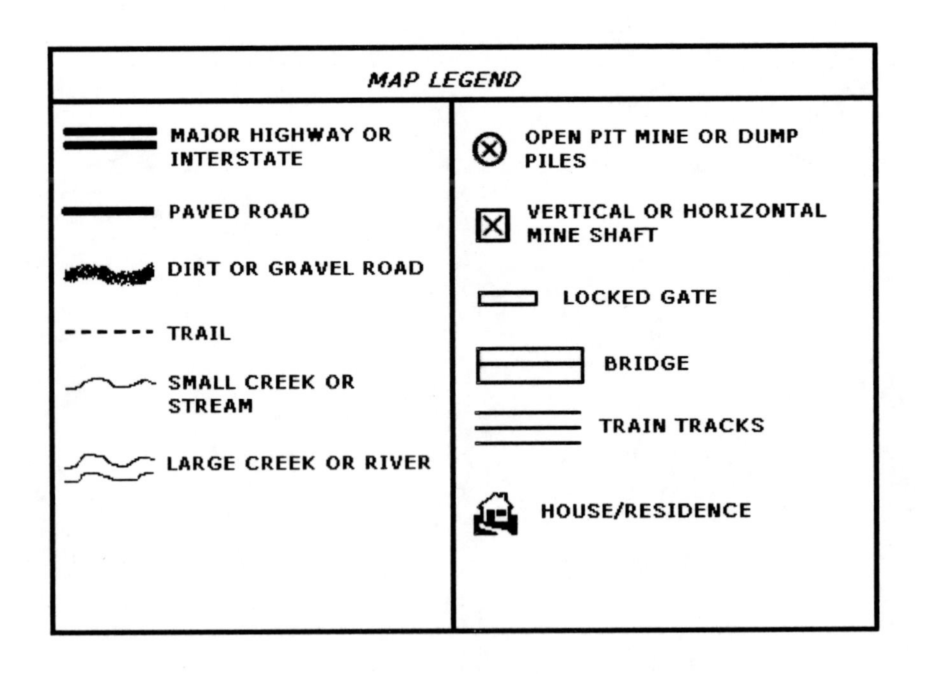
MAP LEGEND
MAJOR HIGHWAY OR INTERSTATE
PAVED ROAD
DIRT OR GRAVEL ROAD
TRAIL
SMALL CREEK OR STREAM
LARGE CREEK OR RIVER
OPEN PIT MINE OR DUMP PILES
VERTICAL OR HORIZONTAL MINE SHAFT
LOCKED GATE
BRIDGE
TRAIN TRACKS
HOUSE/RESIDENCE

SITE 1: LITTLE PINE GARNET MINE

LOCATION: Madison County, North Carolina.

BEST SEASON: Any, weather permitting.

PROPERTY OWNER: Private (Jackie Ball).

MATERIAL TO COLLECT: Large almandine garnet crystals.

TOOLS: 3-lb. sledgehammer, rock chisels, prybar, rock pick, shovel, 1/2" sifting screen, headlamp, lantern.

VEHICLE: Any.

DIRECTIONS: From Asheville, North Carolina, take US Highway 19-23 North to the US 25-70 Marshall exit, turn left going towards Marshall, drive 12.5 miles to Little Pine Road, turn left and drive 4.2 miles to Roberts Branch Road, when you get to Roberts Branch Road you will see a barn with a sign on it that says "GARNET MINE," turn left and drive 0.4 miles to a gravel drive on the left, follow the drive across a small creek and park on the other side in the open parking area.
GPS Coordinates: 35 46.200 N 082 44.260 W

WHAT TO LOOK FOR: You can hike straight up the road (south) about 100 yards and dig and sift for loose garnet crystals in the dump piles on the left (east) side of the road, or hike up the road to the left (east) up the hill approximately 250 yards to the mine at the top of the hill on the right. You will need a headlamp or lantern to explore inside the mine. Once inside, you will see garnet crystals protruding from the walls and ceiling. Most crystals are well formed 12-sided almandine/iron garnets. You may also find twin crystals and nice matrix specimens. The crystals range in size from 1/4" to 4". If you're lucky you might find the rarer gem quality deep red almandine crystal.

I had the privilege of talking with Mr. Jack Ball about his mine in 1993. He told me a little bit about the history of the mine and I will repeat it here as I remember it. He said that the mine was worked off and on since the early 1900s for abrasive garnet and at one time Tiffany's of New York leased the mine from his grandfather for gem quality almandine garnet. He told me a story of how as children he and his friends would take the garnet crystals that were discarded on the dump piles—they were as big as bowling balls but not good enough for gems or abrasive material—the children would roll the garnets down the hill from the mine into the creek for fun!!! Today, the mine is operated by his son, Jackie Ball. You can still find plenty of beautiful cabinet specimens here, and as long as hand tools are all that are used to remove the specimens there should be enough to last another 100 years.

FEE: When you cross the French Broad River on Little Pine Road look to your right and you will see the Davis Grocery store (BP gas station) you will need to stop here before proceeding to the mine, there is a $5.00 per-person per-day fee that you pay at the store.

SAFETY: As with any mine you should wear protective clothing and eye protection when digging, the floor of this mine is on a slant in places and can be slippery which makes it difficult to walk on without proper shoes, I would suggest boots with good tread, this mine is not a place for children.

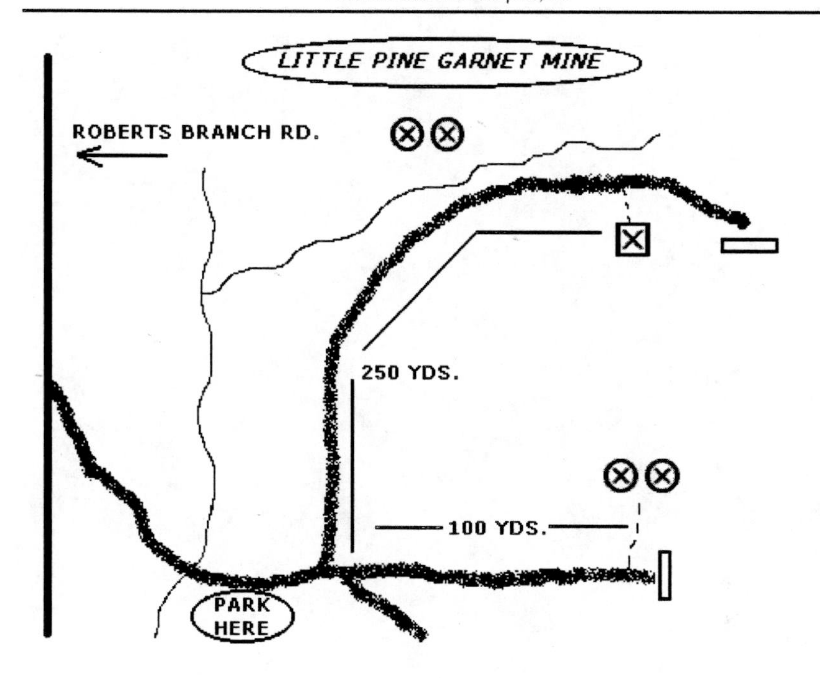
LITTLE PINE GARNET MINE
ROBERTS BRANCH RD.
250 YDS.
100 YDS.
PARK HERE

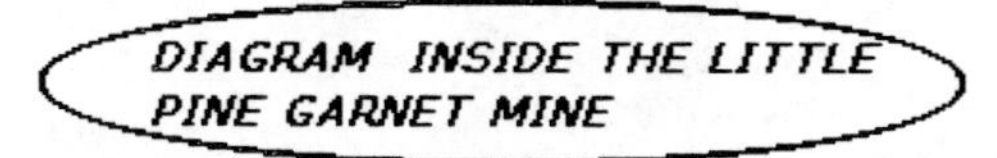
DIAGRAM INSIDE THE LITTLE PINE GARNET MINE

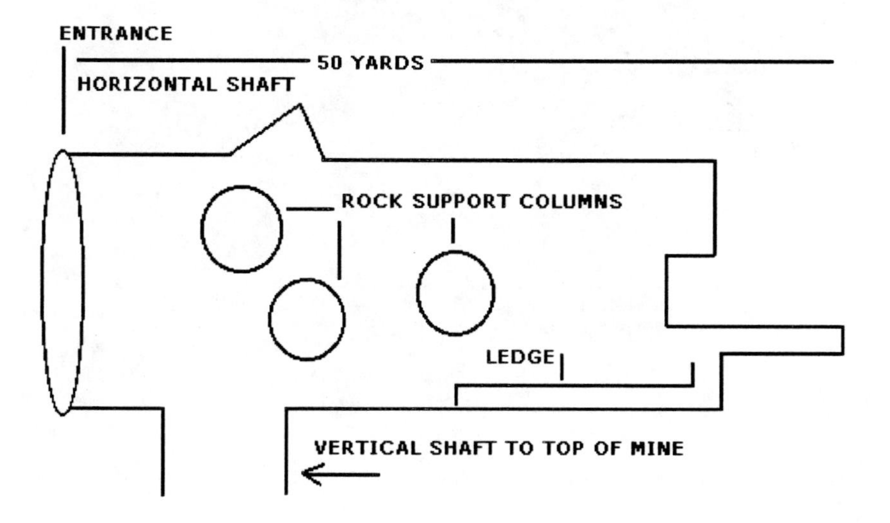
ENTRANCE
50 YARDS
HORIZONTAL SHAFT
ROCK SUPPORT COLUMNS
LEDGE
VERTICAL SHAFT TO TOP OF MINE

Entrance to the Little Pine Garnet Mine

2.5 pound garnet crystal from Little Pine

Inside the Little Pine Mine

Garnet crystal in matrix rock from the Little Pine Mine

SITE 2:
NEWBRIDGE RHODOLITE GARNET

LOCATION: Buncombe County, North Carolina.

BEST SEASON: Any, weather permitting.

PROPERTY OWNER: Private (James Redmond).

MATERIAL TO COLLECT: Rhodolite garnet crystals in schist.

TOOLS: Small rock hammer, flathead screwdriver, rock chisel.

VEHICLE: Any.

DIRECTIONS: From Asheville, North Carolina, take US Highway 19-23 North to the Elk Mountain Road/Woodfin exit, turn right and drive 0.2 miles to Elkwood Avenue, turn left and drive 0.5 miles to Merrimon Avenue, turn left go about 100 yards and turn right into the Newbridge Food Lion parking lot. Park on left side of parking lot next to the vacant field.
GPS Coordinates: 35 38.385 N 082 34.591 W

WHAT TO LOOK FOR: Search the large boulders and rocks in the vacant field. You will find nice small pink rhodolite garnet crystals in the schist rock. You can pluck the crystals out of the rock or look for a more solid matrix piece for display. These crystals range in size up to 1/2". This property is owned by the Redmond family. I have talked with James Redmond and he said it was ok to collect but not to dig any holes. Surface collecting is permitted.

FEE: There is no fee to collect at this site.

SAFETY: This is a safe place to collect, watch children near the parking lot.

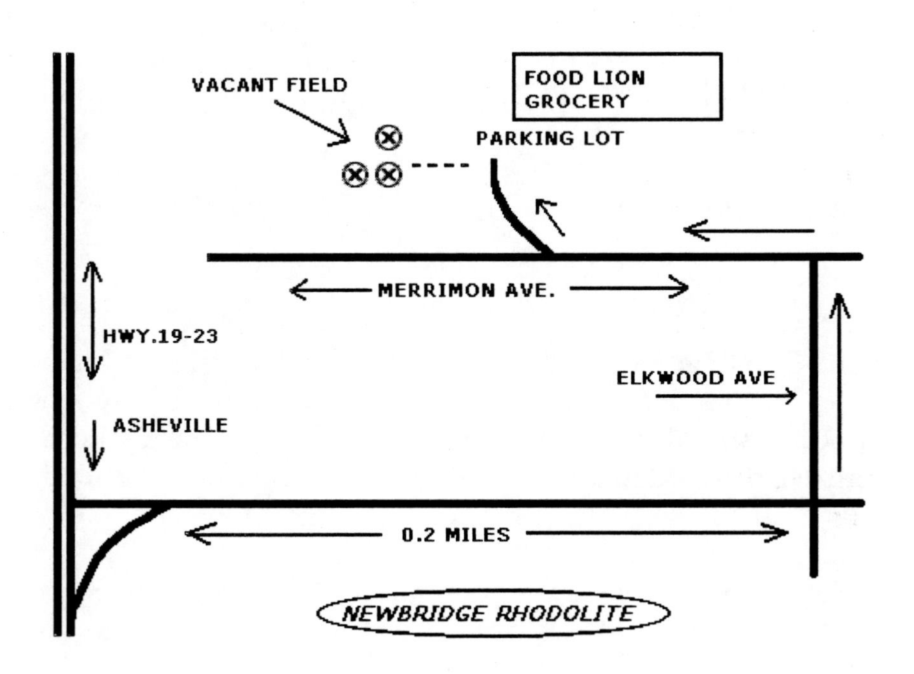

SITE 3: GOLDSMITH MINE

LOCATION: Buncombe County, North Carolina.

BEST SEASON: Any, weather permitting.

PROPERTY OWNER: Private.

MATERIAL TO COLLECT: Olivine, spinel, chalcedony, moonstone.

TOOLS: 3-lb. sledgehammer, rock chisels, rock pick.

VEHICLE: Any.

DIRECTIONS: From Asheville, North Carolina, take US Highway 19-23 North to the Jupiter/Barnardsville 197 exit, turn right towards Barnardsville, drive 3.3 miles on NC 197 to Charcoal Road, turn left, drive 0.2 miles to the mine on the right side of the road.
GPS Coordinates: 35 47.121 N 082 29.892 W

WHAT TO LOOK FOR: Use your sledge to crack open the boulders to find various colors of chalcedony, or dig in the dump piles for moonstone, olivine and small black spinel crystals. This mine was originally mined for mica and feldspar and more recently, vermiculite.

FEE: There is no fee to collect at this mine.

SAFETY: This is a safe place to collect.

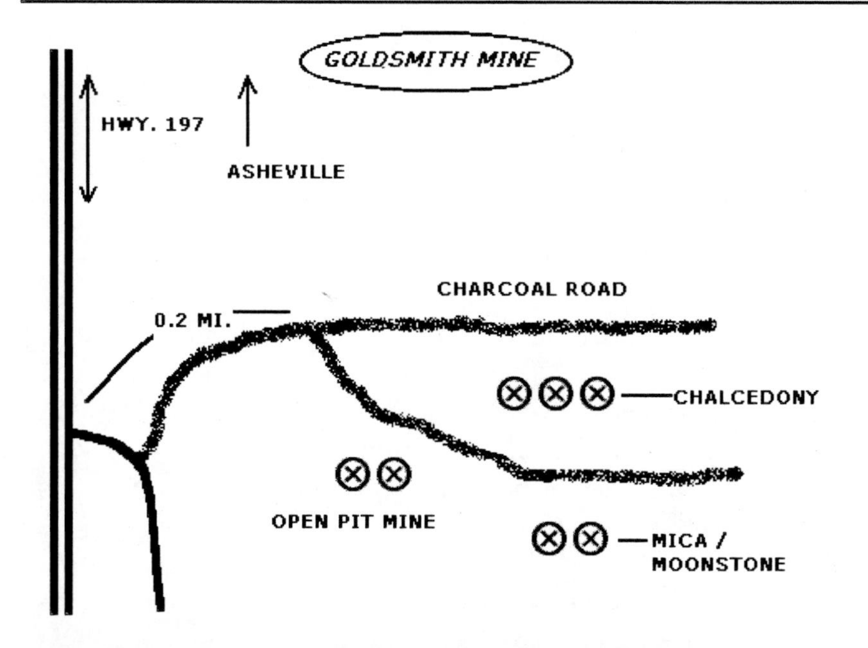

Quartz chalcedony specimen from the Goldsmith Mine

SITE 4:
WALKER CREEK KYANITE

LOCATION: Buncombe County, North Carolina.

BEST SEASON: Any, collecting can be done at any time weather permitting, but in the winter the forest service sometimes closes the gate to the forest road which adds about a 4.0 mile hike to the collecting site.

PROPERTY OWNER: National Forest Service.

MATERIAL TO COLLECT: Large blue kyanite crystals, garnet, schorl (black tourmaline), mica, apatite.

TOOLS: 3-lb. sledgehammer, rock chisels, rock pick, shovel, 1/2" sifting screen.

VEHICLE: Any.

DIRECTIONS: From Asheville, North Carolina, take US Highway 19-23 North to the Jupiter/Barnardsville 197 exit, turn right going towards Barnardsville, drive 6.0 miles on NC 197 into Barnardsville, turn right onto Dillingham Road, drive 4.9 miles to the US Forest Road no. 74 (gravel road), drive 4.4 miles to a pulloff on the left, (Perkins Road no. 175), this is no longer a road but a hiking trail, park here and follow the trail east approximately 750 yards, here you will come to a small stream, continue across the stream on the trail, hike approximately another 900 yards to the head of Walker Creek. From here walk

upstream approximately 30 yards and cross the creek, on the other side you will see a trail leading east up the mountain. From here you want to follow the trail east about 110 degrees if you have a compass (recommended) up the mountain, the trail will run parallel with the creek about 100 yards above the creek, you will follow the trail approximately 850 yards, you will come to the pegmatite of feldspar and quartz. You will see prospecting holes from previous rockhounds. It will take approximately 30-45 minutes to hike up the mountain from the parking area.
GPS Coordinates: 35 45.395 N 082 21.316 W (parking area)
GPS Coordinates: 35 45.212 N 082 20.580 W (collecting site)

WHAT TO LOOK FOR: Here you can dig and sift in the dump piles for nice blue blades of kyanite (some facet and cabbing grade), or you can dig deep and look for the large boulders and crack them with your sledge to find the large blue bladed specimens of kyanite in matrix. You may also find garnet, schorl, and green apatite.

FEE: There is no fee to collect at this site, forest service rules and regulations apply.

SAFETY: This is a remote location in the woods so be aware of wildlife such as bears, mountain lions and snakes to mention a few, this is not a place for small children, the hike would be to hard for a child.

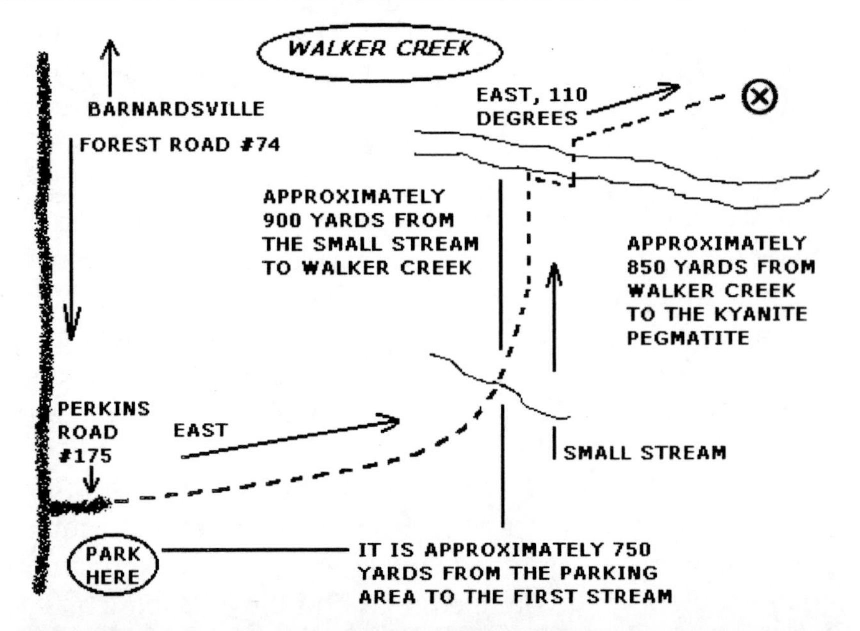

Kyanite in matrix, Walker Creek
The kyanite crystal measures 12" x 2.5" x 1" thick

Dump piles at the Walker Creek collecting site

You will pass this waterfall at the bottom of Walker Creek just before you come to the parking area

SITE 5-A:
RAY MICA MINE

LOCATION: Yancey County, North Carolina.

BEST SEASON: Any, weather permitting.

PROPERTY OWNER: National Forest Service.

MATERIAL TO COLLECT: Beryl crystals (loose and in matrix), fluorescent apatite, garnet, tourmaline (schorl), mica, amazonstone, green tourmaline, thulite.

TOOLS: 3-lb. sledgehammer, 6-lb. sledgehammer, rock chisels, rock pick, 1/2" sifting screen, shovel.

VEHICLE: Any to the forest road, four-wheel drive up the forest road to the mine.

DIRECTIONS: From Asheville, North Carolina, take US Highway 19-23 North to exit 9, Burnsville/Spruce Pine exit, turn right towards Burnsville on US 19 North, drive 17.1 miles into Burnsville, turn right onto Pensacola Road/NC 197 South, drive 0.7 miles to Bolens Creek road, turn left, drive 1.3 miles to Ray Mine Road, turn left follow Ray Mine Road to the dead end at the US Forest Road. If you have a four-wheel drive vehicle you can continue the last 0.3 miles to the collecting site, if not park to the side of the forest service road (do not block the forest road), the walk is about 10 minutes up the road to the collecting area.
GPS Coordinates: 35 53.240 N 082 16.728 W

WHAT TO LOOK FOR: Here you can dig in the dump piles and sift the material in the creek or dry sift to find small beryl crystals, or dig deep to find the larger boulders to crack with your sledgehammer. You can find large beryl crystals in matrix with garnets, mica, amazonstone and schorl (black tourmaline). If you are lucky you may find small green tourmaline crystals in matrix but these are very rare. If you are a fluorescent mineral collector, visit this site at night with a short wave UV light. There is plenty of massive apatite here that fluoresces a bright orange color. You may also find the pink form of zoisite called thulite, this material is also fluorescent. The beryl is a pale green to dark green in color and you may also find some aquamarine. The Ray Mica Mines have been worked for mica since the 1800s. There are still plenty of nice mica specimens here, but today the beryl specimens are what people are looking for. I heard a story of a miner who would bring beryl crystals home to his children—they would use the larger ones in their tree fort as stools to sit on!!!

FEE: There is no fee to collect at this site, follow forest service rules and regulations.

SAFETY: This area is covered with many old vertical mine shafts, some are covered with brush and overgrowth and are hard to see, some of these shafts can be hundreds of feet deep, so be careful while searching the area, this is not a place for small children.

NOTE: If you are heading towards Spruce Pine from this location, stop in the town of Micaville, N.C. on the way and visit the "Old Miners Shack," this is owned by Luther Thomas and is probably the best collection of N.C. minerals and gemstones in the state. Luther is 93 years old and has a wealth of information about mining and mineral collecting locations.

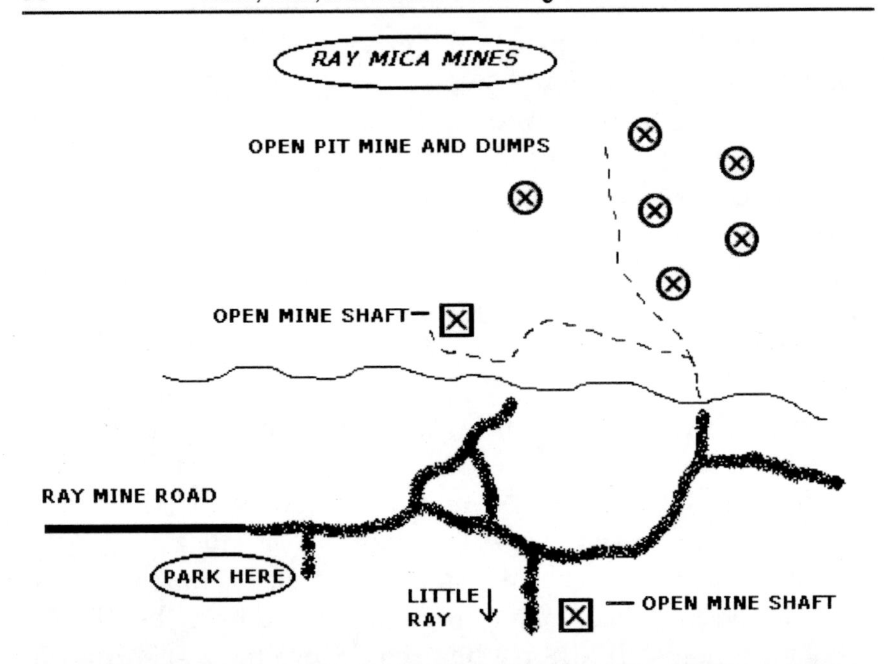

Forest Service sign warning of the open shafts in the area

4" beryl crystal in matrix from the Ray Mine

The black hole in the center of this picture is just one of several vertical mine shafts in the area. Some are hundreds of feet deep so watch your step!

Dig through the dump material to find nice beryl specimens

The specimen on the left is an 11.5 pound section of blue beryl crystal from the Ray Mine

The Old Miner's Shack
This building houses the best collection of North Carolina minerals in the state.

You will need a four-wheel drive vehicle to navigate the rough forest road up to the Ray and Little Ray mine sites

SITE 5-B: LITTLE RAY MICA MINE

LOCATION: Yancey County, North Carolina.

BEST SEASON: Any, weather permitting.

PROPERTY OWNER: National Forest Service.

MATERIAL TO COLLECT: Green, blue, golden beryl crystals, apatite, tourmaline (schorl), mica, amazonstone.

TOOLS: 3-lb. sledgehammer, 6-lb. sledgehammer, rock chisels, rock pick, 1/2" sifting screen, shovel.

VEHICLE: Any to forest road, four-wheel drive up the forest road to the mine.

DIRECTIONS: Refer to map for the exact location of the mine.
GPS Coordinates 35 53.240 N 082 16.728 W

WHAT TO LOOK FOR: I have listed this site as 5-B since it is in the same area as the Ray Mine site. You will find the same material at the Little Ray as the Ray Mine, but at the Little Ray you can also find nice crystals of golden beryl. Collecting technique is the same as the Ray Mine. You can dig and sift for smaller material or break the larger rocks to find specimens in matrix.

FEE: There is no fee to collect at this site, forest service rules apply.

SAFETY: As with the Ray Mine, there are many vertical mine shafts in the area, watch where you walk.

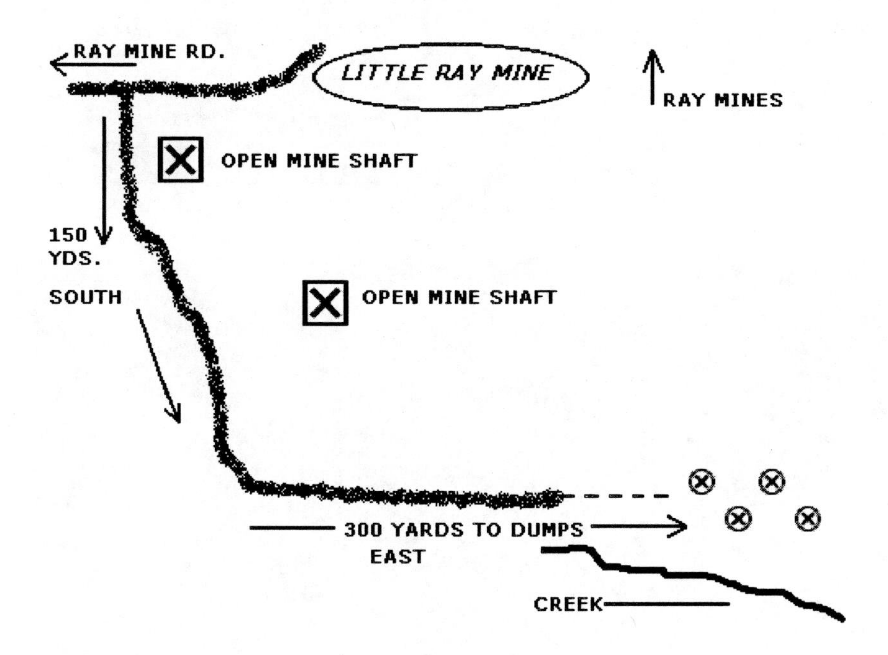

Golden beryl crystal in matrix from the Little Ray Mine

Dump piles at the Little Ray site

This vertical shaft is fenced off at the Little Ray site

SITE 6:
SINKHOLE MINE

LOCATION: Mitchell County, North Carolina.

BEST SEASON: Any, weather permitting.

PROPERTY OWNER: Private (Ed Silver).

MATERIAL TO COLLECT: Green apatite crystals in matrix, small gemmy red almandine garnets, moonstone, pale green amazonstone, kyanite.

TOOLS: 3-lb. sledgehammer, rock chisels, rock pick.

VEHICLE: Any.

DIRECTIONS: From Asheville, North Carolina, take US Highway 19-23 North to exit 9, Burnsville/Spruce Pine exit, turn right going towards Spruce Pine on US 19 North, drive 22.9 miles to Highway NC 80 North, turn left and drive 8.4 miles into the town of Bandana, the mine is on the right side of the road. If the gate is up you can park at the bottom of the mine and walk in, if it is open you can drive in and follow the road 0.1 mile to the top of the mine, there are trespassing signs posted but rock collecting is permitted.
GPS Coordinates: 35 58.534 N 082 10.588 W

WHAT TO LOOK FOR: In the dump piles you can find nice hexagonal light green apatite crystals in a matrix of white feldspar with small gemmy red garnets, the apatite crystals can reach 1" across and 4" long. The garnets are small about 1/4" or smaller. You may also find pale green amazonstone and moonstone, and more rarely blue kyanite. The apatite crystals here will glow a bright orange under a shortwave UV light. It is believed that the Sinkhole Mine was first mined about 300 years ago by the Spaniards and native Indians for silver and mica. Today no silver is found but there are still plenty of nice cabinet specimens to be had.

FEE:

There is no fee to dig here, the owner asks only that you respect the property and no trash is left in the mine.

SAFETY:

If collecting around the bottom of the dumps watch for loose rocks rolling down from the top, if you decide to hike to the top and look around be careful of the steep drop over the edge, there is also a vertical mine shaft (marked on the map) on the property, be sure to stay away from it, keep an eye on small children.

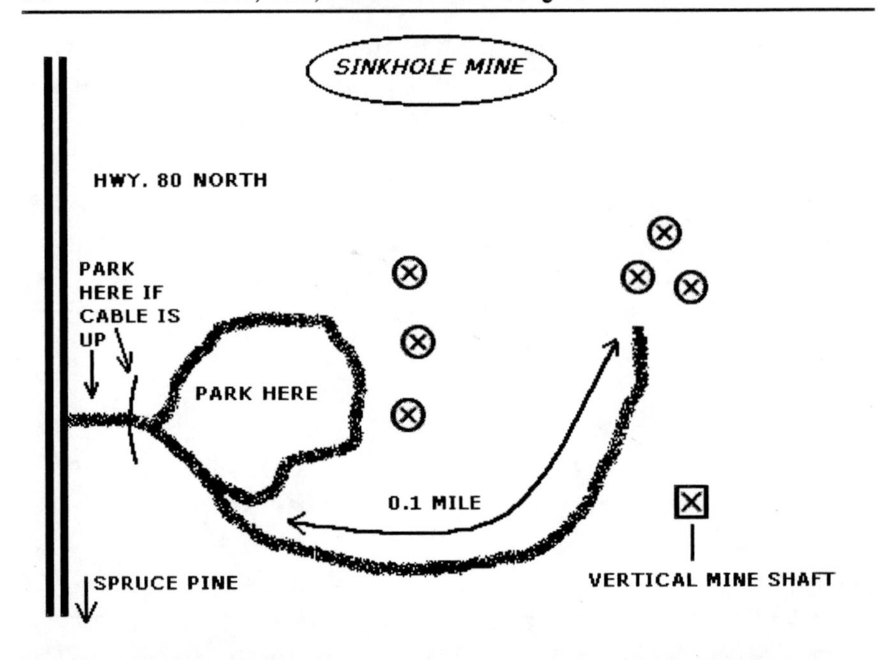

The Sinkhole Mine

3" apatite crystal with garnet crystals in matrix from the Sinkhole Mine

The trash in the bottom of this vertical shaft at the Sinkhole is one of the reasons the owner blocked the entrance for almost two years. Today you can drive in to the site but take whatever you bring in back out with you!

SITE 7-A: ABERNATHY MINE

LOCATION: Mitchell County, North Carolina.

BEST SEASON: Any, weather permitting.

PROPERTY OWNER: Private.

MATERIAL TO COLLECT: Well formed transparent green apatite crystals, gemmy red almandine garnets, mica.

TOOLS: 3-lb. sledgehammer, rock chisels, rock pick, rubber boots, headlamp, lantern.

VEHICLE: Any.

DIRECTIONS: From the Sinkhole Mine continue on NC 80 North for 1.5 miles to Roses Branch Road, turn left, drive 0.8 miles to the train bridge and park here. Walk to the left (south) following the train tracks approximately 2500 yards (just under 1.5 miles), the walk is about 20-30 minutes, you will see the dump piles of white feldspar from the mine on the right side (west) of the tracks and going down the hill to the river.
GPS Coordinates: 35 59.457 N 082 11.448 W (parking area)
GPS Coordinates: 35 58.226 N 082 11.457 W (mine)

WHAT TO LOOK FOR: The dump piles from the Abernathy Mine come down the mountain from the mine and cross the train tracks and continue to the river. There are tons of material to search through. You will find beautiful green crystals of apatite in matrix. This apatite is a fluorescent orange under a shortwave UV light. You can also find nice red almandine garnets up to 1" in size. If you want to visit the mine, you can either climb the hill across from the tracks to the big shaft at the top—it is a steep climb but it's worth it to see the old mine—or you can walk to the south end of the dump piles on the tracks (about 75 yards) and look to the left (east) at the rock wall. You will see a small opening in the rock, about 4 feet wide, go through the opening into the old underground shafts. You will need your rubber boots in here and a light or headlamp. These tunnels go hundreds of feet underground, so be careful.

FEE: There is no fee to collect at the location listed here, the dump piles are on state and railroad property. If you climb the hill to the big mine shaft at the top, this is on private property.

SAFETY: This is not a good place to bring small children, the trains run about every 20-30 minutes in both directions on the tracks that you follow to the mine, and the hill going down to the river is steep.

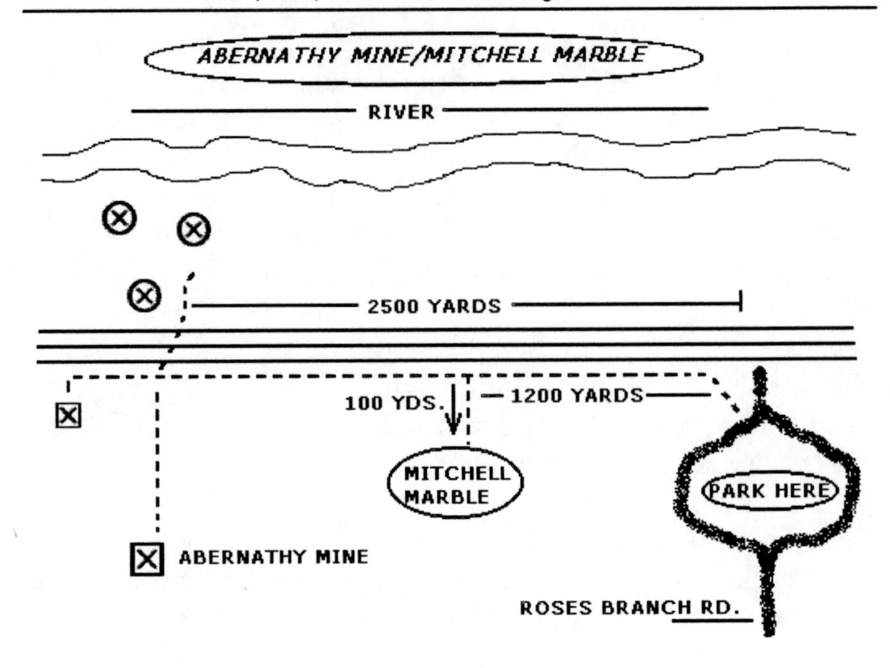
ABERNATHY MINE/MITCHELL MARBLE
RIVER
2500 YARDS
100 YDS.
1200 YARDS
MITCHELL MARBLE
PARK HERE
ABERNATHY MINE
ROSES BRANCH RD.

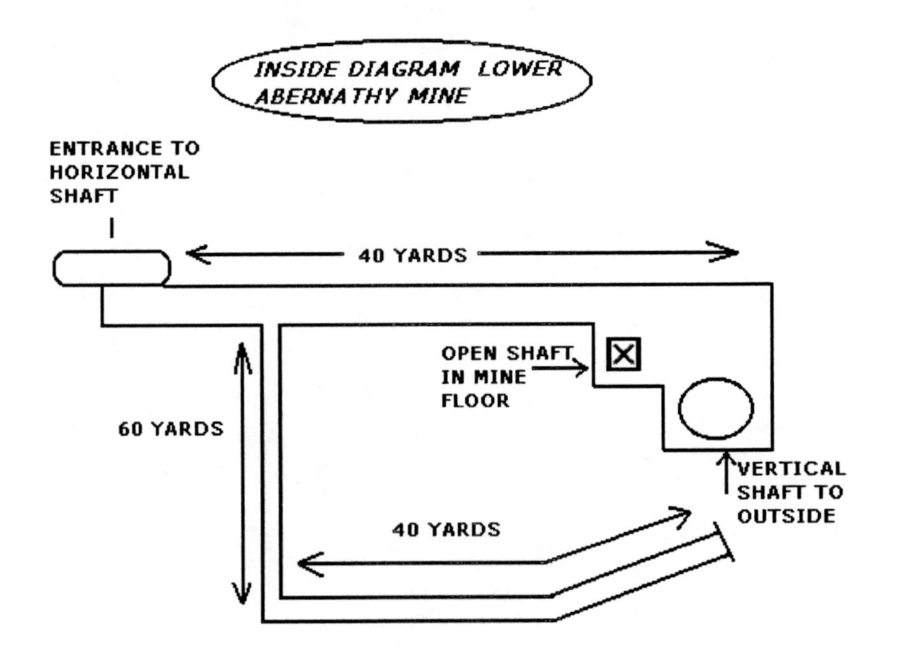
INSIDE DIAGRAM LOWER ABERNATHY MINE
ENTRANCE TO HORIZONTAL SHAFT
40 YARDS
OPEN SHAFT IN MINE FLOOR
60 YARDS
VERTICAL SHAFT TO OUTSIDE
40 YARDS

Be careful of trains while collecting near the dump piles next to the tracks at the Abernathy Mine

Garnet in matrix from the Abernathy Mine

This small opening in the rock leads into the old shafts at the bottom of the Abernathy Mine

Specimens of garnet in matrix like this piece can be found at the Abernathy Mine. This specimen is in the Smithsonian Museum in Washington, DC.

SITE 7-B: MITCHELL MARBLE

LOCATION: Mitchell County, North Carolina.

BEST SEASON: Any, weather permitting.

PROPERTY OWNER: Private.

MATERIAL TO COLLECT: White marble.

TOOLS: 3-lb. sledgehammer, large sledge hammer, rock chisels.

VEHICLE: Any.

DIRECTIONS: Refer to Abernathy Mine/Mitchell Marble map for exact location of site.

WHAT TO LOOK FOR: I have included this site as 7-B. It is along the train tracks going to the Abernathy Mine. Here you can find nice boulders and pieces of white marble.

FEE: There is no fee to collect at this site.

SAFETY: This is a safe place to collect, watch for trains when walking to the site.

SITE 8: CRABTREE EMERALD MINE

LOCATION: Mitchell County, North Carolina.

BEST SEASON: Any, weather permitting.

PROPERTY OWNER: Private (Terry Ledford).

MATERIAL TO COLLECT: Emerald crystals in matrix, beryl, garnet, tourmaline (schorl).

TOOLS: 3-lb. sledgehammer, large sledgehammer, rock chisels, rock pick, 1/4" sifting screen, shovel.

VEHICLE: Any.

DIRECTIONS: From Asheville, North Carolina, take US Highway 19-23 North to exit 9, Burnsville/Spruce Pine exit, US 19 North, turn right towards Spruce Pine, drive 25.4 miles to Crabtree Road, turn right, drive 4.7 miles to McKinney Mine Road, turn left, drive 2.3 miles to Chestnut Grove Road, turn left, drive 1.0 miles to Emerald Mine Road, this road will turn to gravel, follow it 1.2 miles to the mine on the left.
GPS Coordinates: 35 52.477 N 082 07.197 W

WHAT TO LOOK FOR: Here you can find loose emerald crystals by digging and sifting the dump material, or you can pound on the huge emerald-bearing boulders around the dumps. There are orange/red garnets here and some white and pale green beryl specimens. You will find plenty of feldspar/quartz matrix specimens with nice black tourmaline crystals and emeralds together. Many people cut cabachons from the matrix pieces with the emerald in the feldspar/quartz. This mine has been worked for emerald since the 1890s. In the 1980s the mine had to be shut down for good. Today no commercial mining takes place, but there are still plenty of huge boulders and dump piles to be searched for nice specimens.

FEE: There is no fee to collect at this mine.

SAFETY: This is a safe area to collect, make sure not to fall into the cold water filled mine.

NOTE:
You will pass the Emerald Village Mine on the way to the Crabtree Mine, it is on McKinney Mine Road, if you have time you should stop here and tour the North Carolina Mining Museum and take the tour of the underground mine, it is worth the stop.

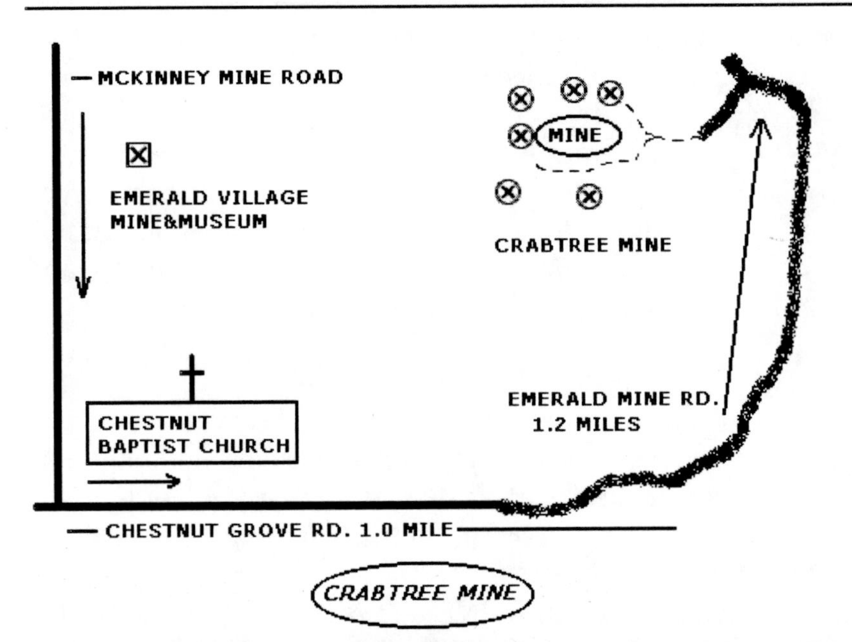

This small pond was once the 200-foot deep Crabtree Emerald Mine shaft

This 84-pound specimen of emeralds, garnet and tourmaline from the Crabtree Mine is displayed in the author's collection

An old piece of minig equipment rusting away at the Crabtree Mine

The Emerald Village Mine
It is worth stopping to take the tour of this historic mine and museum

SITE 9: CHALK MOUNTAIN

LOCATION: Mitchell County, North Carolina.

BEST SEASON: Any, weather permitting.

PROPERTY OWNER: Private (Feldspar Corporation).

MATERIAL TO COLLECT: Hyalite opal, autunite, torbernite.

TOOLS: 3-lb. sledgehammer, rock chisels, rock pick, short wave UV light.

VEHICLE: Any.

DIRECTIONS: From Asheville, North Carolina, take US Highway 19-23 North to exit 9, Burnsville/Spruce Pine exit, turn right towards Spruce Pine on US 19 North, drive 28.3 miles to the entrance of the Chalk Mountain mine on the right.
GPS Coordinates: 35 54.449 N 082 05.979 W

WHAT TO LOOK FOR: This mine produces the most highly fluorescent hyalite opal in the world. The opal forms as a bright green coating on the feldspar rocks that can be easily seen in the daytime. You can also find nice small dark green tabular crystals of torbernite here and bright yellow/green tabular crystals of autunite. If you come at night when the mine is closed with a shortwave UV light you will have no problem finding material. Chalk Mountain is an active mine that produces feldspar, quartz, and mica used for ceramics, light bulbs, etc.

FEE: There is no fee to collect at this mine but collecting is not permitted during working hours, to arrange day and night collecting tours of the mine you need to contact either the Mitchell County Chamber of Commerce or Alex Glover. He is the head geologist for the mine and the president of the chamber of commerce.

SAFETY: This is a good place to collect for adults and children, falling rock can be a hazard so you should stay away from the mine walls, there is plenty of material in the dump piles to go through.

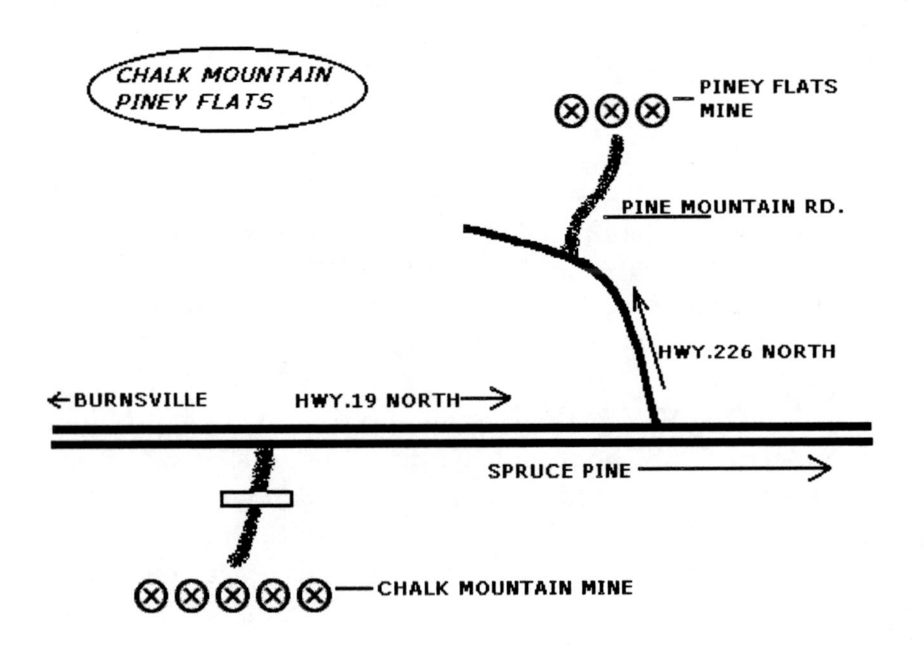

Chalk Mountain as seen from Highway 19 North

The top of Chalk Mountain

SITE 10: PINEY FLATS MINE

LOCATION: Mitchell County, North Carolina.

BEST SEASON: Any, weather permitting.

PROPERTY OWNER: Private (Unimin Corporation).

MATERIAL TO COLLECT: Blue and green hyalite opal, autunite.

TOOLS: 3-lb. sledgehammer, rock chisels, rock pick, short wave UV light.

VEHICLE: Any.

DIRECTIONS: From Asheville, North Carolina, take US Highway 19-23 North to exit 9, Burnsville/Spruce Pine exit, go right towards Spruce Pine on US 19 North, drive 29.3 miles to NC Highway 226 North, turn left and drive 1.7 miles to Pine Mountain Road (when I was last at this location there was no street sign for Pine Mountain Road so watch your odometer), turn right on Pine Mountain Road and drive 0.6 miles to the entrance to the mine, park on the side road to the right and walk into the mine.
GPS Coordinates: 35 56.312 N 082 05.232 W

WHAT TO LOOK FOR: The lower section of this mine is no longer being worked and is not posted. I have collected here for years. You can find nice green hyalite opal and the rarer blue hyalite. There is plenty of bright yellow/green autunite, also. I like to come at night with a shortwave UV light to find the best material. Piney Flats is an active quarry that produces the same type of material as the Chalk Mountain site.

FEE: There is no fee to collect at this mine.

SAFETY: This is a safe mine to collect at if you stay away from the walls, there are many loose rocks and boulders that can be a hazard.

NOTE:
After you leave the Piney Flats Mine, make your way back to highway 19 North, turn left and drive 0.7 miles to highway 226 South, turn right and drive 4.7 miles to the "Museum of North Carolina Minerals" on the right, next to the parkway, they have some nice specimens on display here.

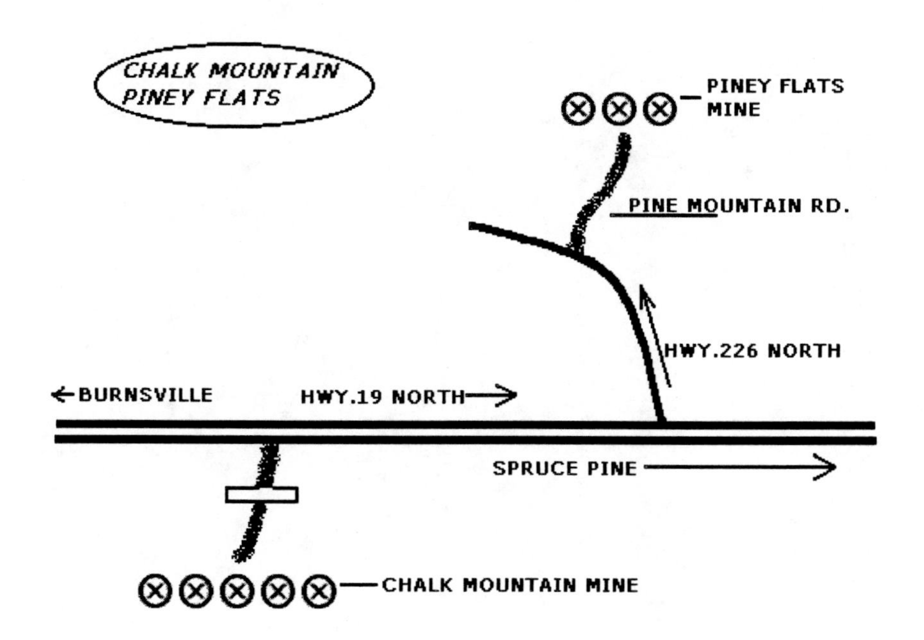

Piney Flats Mine

A view into the quarry at the Piney Flats site

The Museum of North Carolina Minerals in Spruce Pine

SITE 11:
HENSON CREEK AQUAMARINE

LOCATION: Mitchell County, North Carolina.

BEST SEASON: Any, weather permitting.

PROPERTY OWNER: Private (Gary Ledford).

MATERIAL TO COLLECT: Green and blue beryl, aquamarine.

TOOLS: Shovel, 1/2" sifting screen, 3-lb. sledgehammer, rock chisels, rock pick.

VEHICLE: Any.

DIRECTIONS: From the Museum of North Carolina Minerals, return to US 19 North, (intersection of 19 North and 226 South), turn right and drive 10.9 miles to Henson Creek Road (SR 1126), turn left and drive 2.3 miles to Red Dirt Road, turn right and drive 0.2 miles to the mine on the left side of the road, there is a small steel bridge that crosses the creek, cross the bridge and park on the other side at the bottom of the dump piles.
GPS Coordinates: 36 02.672 N 082 03.125 W

WHAT TO LOOK FOR: Here you can dig and sift the dump piles for small pieces and crystals of blue and green beryl and some aquamarine. You can also break the feldspar/quartz rock to find beryl crystals in matrix. The largest crystals in this mine are in a vein about 20 feet deep. This mine has been mined for mica, white china clay and aquamarine in the past.

FEE: There is no fee to collect at this mine.

SAFETY: This is a safe mine to collect at as long as you do not try to climb on the side walls of the mine.

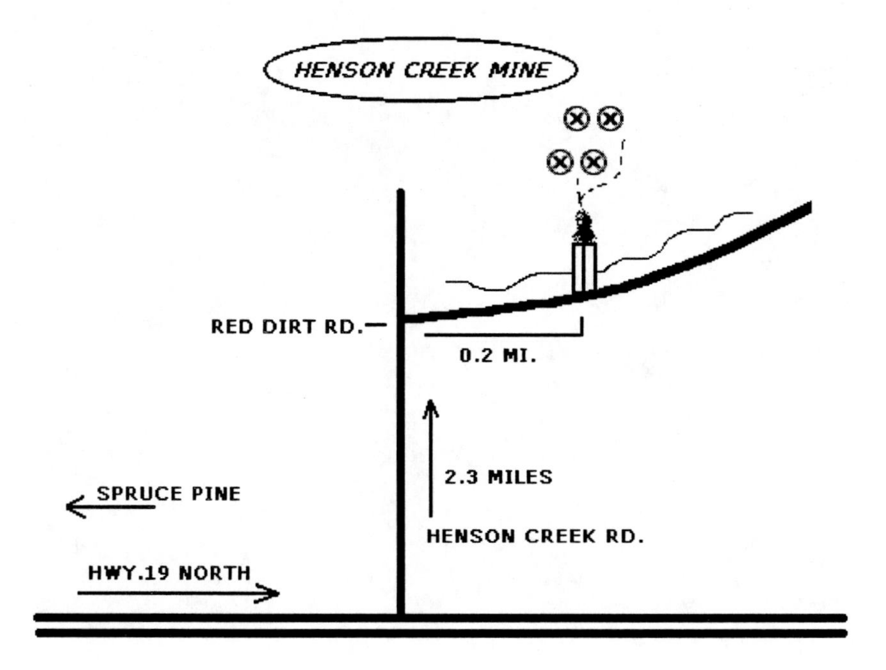

4.5 pound, 6" x 4" blue beryl crystal from the Henson Creek Mine

The Henson Creek collecting site

SITE 12:
WOODS CREEK SAPPHIRE MINE

LOCATION: Haywood County, North Carolina.

BEST SEASON: Any, site is open year round.

PROPERTY OWNER: Private (Woods family).

MATERIAL TO COLLECT: Blue, purple, black, gray, sapphire/corundum.

TOOLS: Shovel, 1/4" sifting screen, rock hammer, wading boots.

VEHICLE: Any.

DIRECTIONS: From Asheville, North Carolina, take Interstate 40 West to exit 33 Leicester, Newfound Road, turn right and drive 0.8 miles to North Hominy Road (SR 1606), turn left and drive 0.6 miles to Willis Cove Road, turn left and drive 0.5 miles to Pressley Mine Road, turn left and follow signs to Woods Creek on the right.

WHAT TO LOOK FOR: The Woods Creek site is an alluvial corundum deposit in the creekbed that runs the length of the property. Minor digging took place in the late 70s and early 80s, in 1987 a local rockhound (Bruce Caminiti) leased the property from Tom Woods and began mining the creek bed. He discovered hundreds of pounds of corundum and the Woods

Creek mine was opened. Today the mine is operated by Branson Woods, who is very helpful to newcomers and will show you the best places to collect. You are allowed to dig and sift anywhere in or around the creek bed. You want to dig down to the gravel layer and sift the material in the creek. You may find sapphires from 1/2" in size up to several pounds. I have seen one piece that weighed over 13 pounds! If you have two or more days you can start a new prospect hole, if you only have a day, there are plenty of holes already started that you can continue to expand.

FEE: The fee here is $15.00 per-person per-day, you pay the owner Branson Woods at the mine.

SAFETY: This is a safe place to collect, if you are mining one of the deep holes be careful not to tunnel into the sides, the ground is unstable and will cave in on top of you.

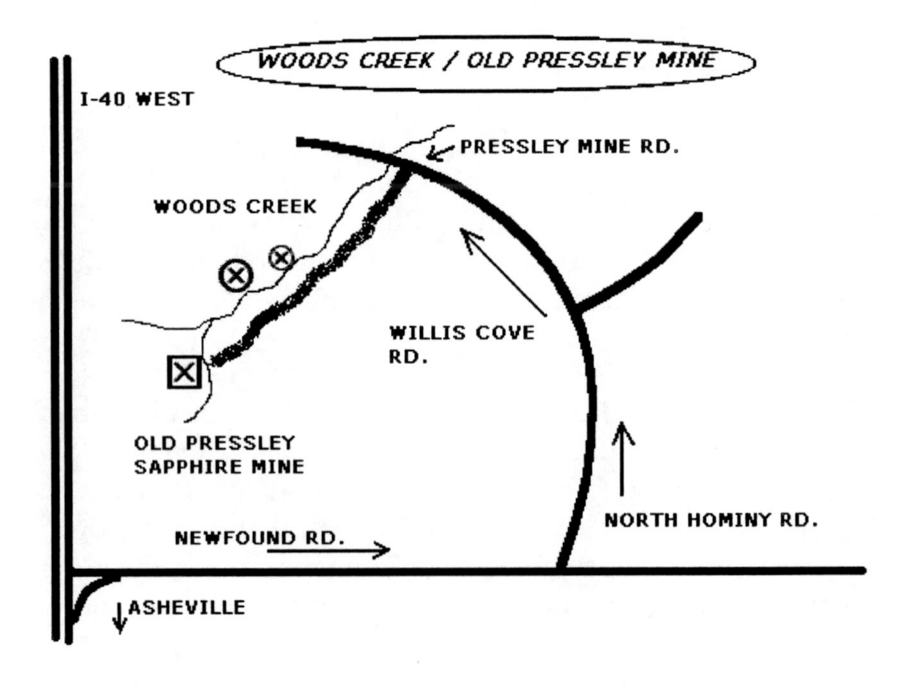

Local rockhounds collecting at Woods Creek

Polished double hex crystal collected at the Woods Creek site

SITE 13:
OLD PRESSLEY SAPPHIRE MINE

LOCATION: Haywood County, North Carolina.

BEST SEASON: Any, weather permitting.

PROPERTY OWNER: Private (George and Brenda McCannon).

MATERIAL TO COLLECT: Blue, purple, black, gray, sapphire/corundum, zircon.

TOOLS: Equipment is provided.

VEHICLE: Any.

DIRECTIONS: When you leave the Woods Creek Sapphire Mine, turn right and continue up the hill to the dead end at the Old Pressley Mine.

WHAT TO LOOK FOR: This is a tourist mine which provides you with digging equipment and a shaded flume to wash your material in. You can purchase a guaranteed bucket of dirt with sapphire inside or pay a $20.00 all day fee to dig your own material in the dump piles across from the flume. This is not a salted mine as many tourist mines are, you will only find native sapphires from this mine. In the past, this mine produced the 1,445 carat "Star of the Carolinas" which still holds the record as the largest star sapphire according to the *Guiness Book of World Records*. It also produced the 1,035 carat "Southern Star."

FEE: Fee varies.

SAFETY: This is a safe place to collect for the whole family.

3-pound 10 ounce sapphire crystal from the Old Pressley Sapphire Mine

SITE 14:
REDMOND MINE

LOCATION: Haywood County, North Carolina.

BEST SEASON: Any, weather permitting.

PROPERTY OWNER: National Forest Service.

MATERIAL TO COLLECT: Azurite crystals as coating on matrix, malachite, cerussite, pyromorphite.

TOOLS: 3-lb. sledgehammer, rock chisels, prybar, rock pick, waders, headlamp, lantern.

VEHICLE: Any in spring, summer and fall, in the winter the road can be rough and a four-wheel drive vehicle should be used.

DIRECTIONS: From Asheville, North Carolina, take Interstate 40 West to Haywood County, drive to the Fines Creek exit, exit 15, set your odometer at exit 15 and continue west on I-40 for 2.3 miles, turn right onto a state maintenance road, follow this road, stay to the right 1.3 miles to the dump piles from the mine on the right side of the road, park here.
GPS Coordinates: 35 40.888 N 083 00.924 W

WHAT TO LOOK FOR: You can dig through the dump piles and find some material but they have been worked quite a bit and it's getting hard to find good specimens. The best material is in the old mine. Walk across the road and follow the trail about 100 feet, you will see the entrance to the shaft. You will need waders and a headlamp or lantern to enter the mine. It has water about 2-3 feet deep for the first 30 feet or so, then it is dry. The walls and ceiling contain nice specimens of dark blue azurite, green malachite and clear and white cerussite crystals up to 1/2" long. This is not cutting material but they make excellent cabinet specimens. If you like microminerals, go out of the mine and follow the trail up the hill to the left of the mine about 100 yards. You will see an area where rockhounds have been breaking rocks and boulders. You will find small radiating green crystals of pyromorphite in the rocks. This mine was originally prospected for copper and lead minerals.

FEE: There is no fee to collect at this mine.

SAFETY: This is a safe mine to collect in, the shaft seems to be very stable but I would leave the children at home.

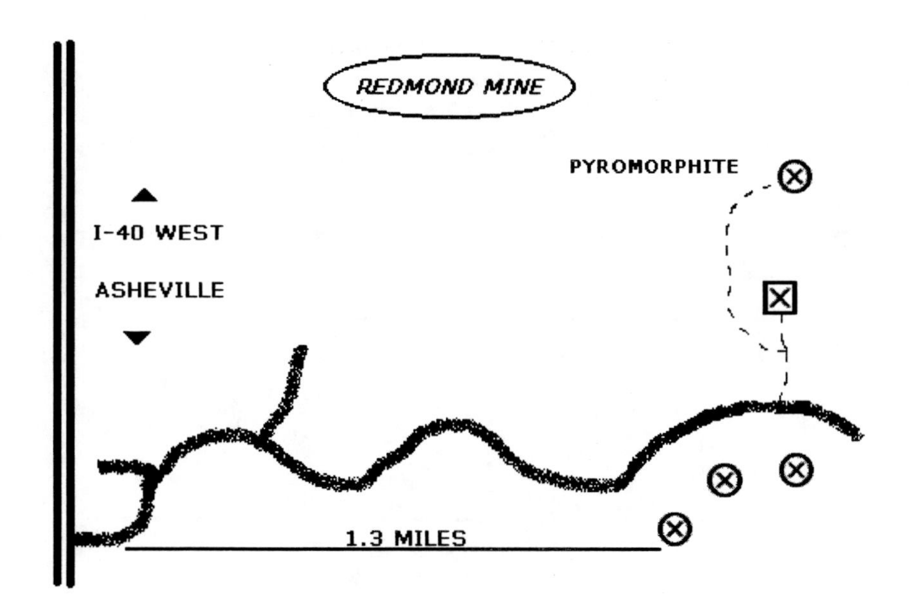

The author's son, R.J., inspects a mineral specimen at the entrance to the Redmond Mine

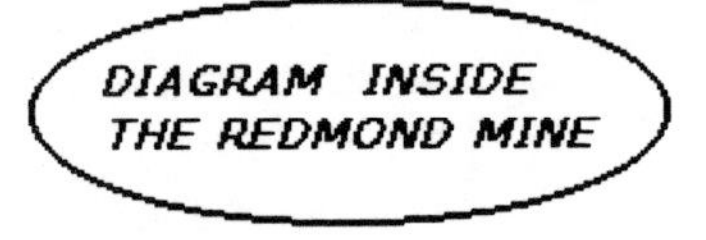

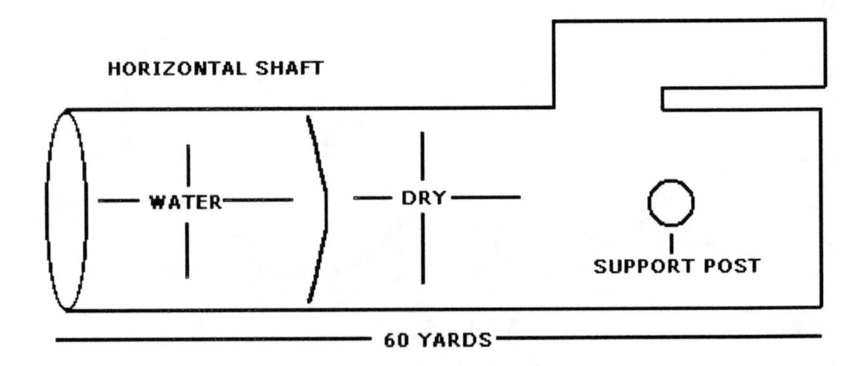

Franklin Area Mines

Mason's Ruby and Sapphire Mine

Cherokee Ruby Mine

Mason Mountain Rhodolite

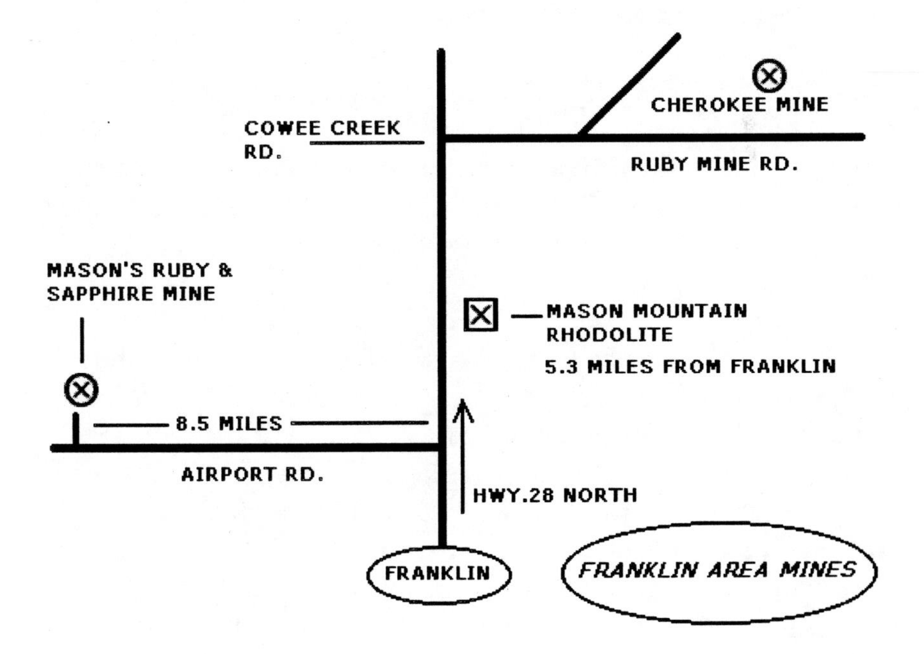

SITE 15:
MASON'S RUBY & SAPPHIRE MINE

LOCATION: Macon County, North Carolina.

BEST SEASON: Open April 1–October 31.

PROPERTY OWNER: Private (Coralee Campbell).

MATERIAL TO COLLECT: Rubies, sapphires, garnets.

TOOLS: Tools are provided by the mine. If you bring your own tools, I would suggest a shovel, bucket, and a 1/4" sifting screen.

VEHICLE: Any.

DIRECTIONS: From Asheville, North Carolina, take Interstate 40 West to exit 27, Waynesville/Cherokee/19-23-74, follow to exit 103, bear left onto US Highway 23-74 Waynesville/Sylva, follow to exit 81, Dillsboro/Franklin/Atlanta, US Highway 23-441 South, follow to the town of Franklin, turn right onto 441 Business (Main Street), follow through town to NC 28 North, turn right and drive 2.7 miles to Airport Road, turn left, drive 8.5 miles to Upper Burningtown Road (you will see the Mt.Sini Church) turn right and drive 1.0 mile to the mine. Address 6945 Upper Burningtown Road.

WHAT TO LOOK FOR: This is one of the better locations in the Franklin area. It has recently come under new ownership. This mine is not salted or enriched with non-native stones. You will find rubies, sapphires, and some garnets. For a small fee you are allowed to dig in the dump piles from the mine. All equipment is provided or you can bring your own. The mine also provides a shaded flume for you to screen your material.

FEE: $10.00 per-person per-day, covers all day digging.

SAFETY: This is a safe place to collect for the whole family.

NOTE:
When you come to the town of Franklin, drive into town on 441 Business (Main Street), look on the left for (Ruby City), this is a gem and mineral store and they have an excellent free museum with some unique and unusual specimens. When you leave here, continue up Main Street to Phillips Street, turn left and then immediately right into the parking lot for the Old Jail Museum, this is also a nice museum with a nice collection of local minerals and gems plus more.

SITE 16:
CHEROKEE RUBY MINE

LOCATION: Macon County, North Carolina.

BEST SEASON: Open April 1 through October 31, 9 A.M. to 4 P.M., closed Wednesday.

PROPERTY OWNER: Private (Effie McCrackine).

MATERIAL TO COLLECT: Rubies, sapphires, rutile, moonstone, silliminite, garnet (rhodolite and pyrope).

TOOLS: Equipment is provided by the mine.

VEHICLE: Any.

DIRECTIONS: From Franklin, North Carolina, take NC Highway 28 North 6.4 miles, turn right onto Cowee Creek Road (SR 1340), drive 1.4 miles to a fork in the road, take the right fork onto Ruby Mine Road, drive 2.5 miles to the mine on the left.

WHAT TO LOOK FOR: This mine is one of the few unsalted mines with all native stones in the Cowee Valley. This mine produced the largest pure ruby in the US, it is certified with the GIA (Gemological Institute of America) and weighs 1,070 carats. The owner Mrs. McCrackine is very friendly and helpful to all visitors. To add a piece of ruby or sapphire from the famous Cowee Valley to your collection, you should definitely visit this mine.

FEE: The fee is $6.00 for adults and $4.00 for children 12 and under which covers all day collecting.

SAFETY: This is a safe place to collect for the whole family.

The sign at the entrance of the Cherokee Ruby Mine
Note the "NOT SALTED"

SITE 17:
MASON MOUNTAIN RHODOLITE

LOCATION: Macon County, North Carolina.

BEST SEASON: Open March 15 through November 1.

PROPERTY OWNER: Private (Brown and Martha Johnson).

MATERIAL TO COLLECT: Rhodolite garnet, kyanite, smoky quartz.

TOOLS: Equipment is provided by the mine, if you get to collect in the mine itself I would bring your own tools, I would suggest: rock pick, rock hammer, 3-lb. sledgehammer, rock chisels.

VEHICLE: Any.

DIRECTIONS: From Franklin, North Carolina, take NC 28 North 5.3 miles to the mine on the right side of 28 North (Bryson City Road).

WHAT TO LOOK FOR: This is a salted mine and I would not normally recommend it as a good collecting site, but it does have some historical significance so I decided to include it in this book. This is the mine where the rhodolite garnet was first discovered in 1895, and the owner of the mine still works the vein for rhodolite garnet. Group trips can be arranged with the owner to collect in the actual mine. This is a tourist mine where you can purchase a salted bucket of dirt with non-native material and screen it in the provided flume. The minerals listed above are the only native stones found here. If you can arrange to collect in the mine itself, it would be worth a trip here to add some of this rhodolite to your collection.

FEE: Varies.

SAFETY: This is a safe place to collect.

30-pound boulder of rhodolite garnets in matrix from the Mason Mountain Mine

SITE 18: CORUNDUM HILL

LOCATION: Macon County, North Carolina.

BEST SEASON: Any, weather permitting.

PROPERTY OWNER: State of North Carolina.

MATERIAL TO COLLECT: Corundum, ruby, sapphire.

TOOLS: 1/4" sifting screen, shovel.

VEHICLE: Any.

DIRECTIONS: From Asheville, North Carolina, take Interstate 40 West to exit 27, Waynesville/Cherokee/19-23-74, follow to exit 103, bear left onto US Highway 23-74 Waynesville/Sylva, follow to exit 81, Dillsboro/Franklin/Atlanta, US Highway 23-441 South, follow to the town of Franklin, take the Highlands exit, US Highway 64 East-28 South, drive 5.2 miles to Peaceful Cove Road (the Cullasaja River will be on your right), turn right, park on the other side of the bridge. Follow the river east upstream about 300 yards, there will be a stream that flows into the river from up on the mountain where the Corundum Hill Mine is located, dig and sift in the river.
GPS Coordinates: 35 08.457 N 083 17.693 W

WHAT TO LOOK FOR: The material you find in the Cullasaja River is alluvial corundum that has washed down from the

Corundum Hill Mine. There is a wide variety of color and some hexagonal crystals. The Corundum Hill Mine was the first gem mine in North Carolina. It was opened by C.E. Jenks in 1871. He was looking for gem quality corundum and although he found beautiful gem rubies and sapphires the mine did not produce enough material to be profitable and was eventually closed. Today the mine is on private property and no collecting is allowed, but the location I have listed here will provide you with nice material from the original mine.

FEE: There is no fee to collect at this site.

SAFETY: This is a safe place to collect, be careful with small children in the current of the river.

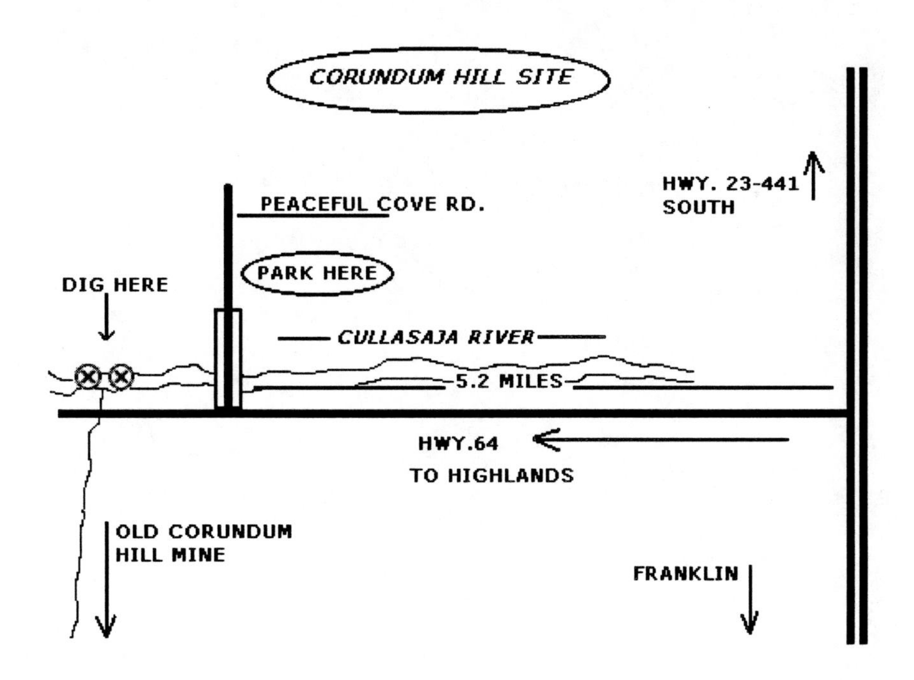

Rockhound Bill Mintz at the entrance of the shaft at the old Corundum Hill Mine site

SITE 19:
BUCK CREEK GARNET

LOCATION: Clay County, North Carolina.

BEST SEASON: Any, weather permitting.

PROPERTY OWNER: National Forest Service.

MATERIAL TO COLLECT: Small gemmy red almandine garnet crystals.

TOOLS: 1/8" sifting screen, shovel, rubber boots.

VEHICLE: Any.

DIRECTIONS: From Franklin, N.C., take Highway 441 South/ 23 South, 441/23 South will turn into US Highway 64 West towards Murphy, stay on 64 West, this will turn into a two lane road over the mountain, continue until you come to the Macon–Clay County line, from the county line drive 2.1 miles to Buck Creek Road on the right, turn right onto Buck Creek Road (road will turn to gravel) drive 0.7 miles to a bridge over the creek, park out of the roadway near the bridge.
GPS Coordinates: 35 05.019 N 083 36.759 W

WHAT TO LOOK FOR: Park near the bridge on Buck Creek Road, dig and sift the sand/gravel in the creek. You will find small 1/8" to 1/4" gemmy red almandine garnet crystals.

FEE: There is no fee to collect here, forest service rules apply.

SAFETY: This is a safe place to collect, watch small children near the creek.

Park near this bridge on Buck Creek Road and dig in the creek to find gemmy red garnets

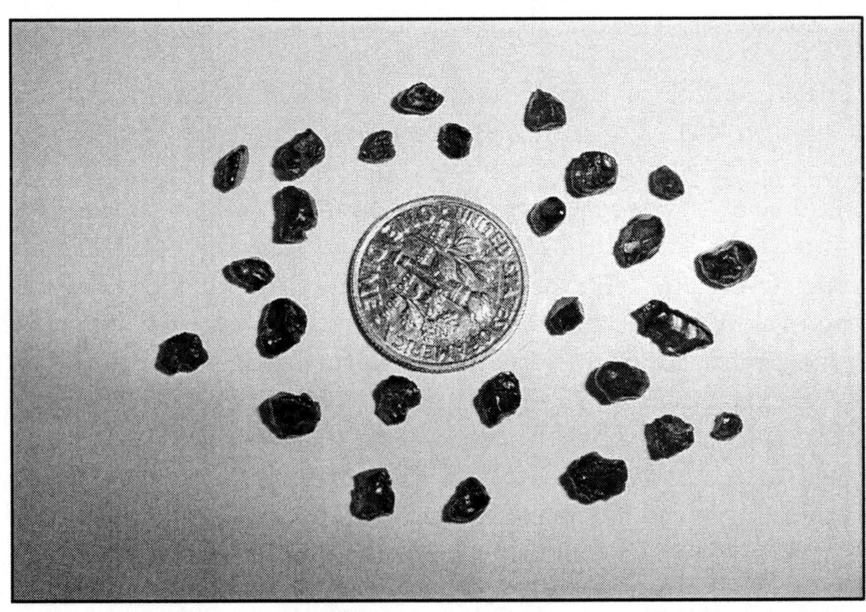

Gem red almandine garnets from Buck Creek

SITE 20: BUCK CREEK, OLD RUBY MINE

LOCATION: Clay County, North Carolina.

BEST SEASON: Any, weather permitting.

PROPERTY OWNER: National Forest Service.

MATERIAL TO COLLECT: Corundum, ruby, blue and gray corundum.

TOOLS: Rock pick, 1/2" sifting screen, shovel, rubber boots/waders.

VEHICLE: Any.

DIRECTIONS: From the Buck Creek Garnet collecting site, continue on Buck Creek Road from the bridge 0.3 miles to a pulloff into a campsite on the right, park at the campsite.
GPS Coordinates: 35 05.071 N 083 36.969 W

WHAT TO LOOK FOR: Starting at the campsite, hike southeast up the creek approximately 50 yards. You will see an old road on the other side. Following the dump piles and trail to the left of the old road, hike about 100 yards northeast up to the old mine shaft. When the mine was operating, the corundum was loaded onto wagons and hauled down the mountain and across the creek to Buck Creek Road and then transported to the processing plant. Sometimes pieces of corundum would fall from the wagon into the creek. Dig and sift the creek bed and you will find a variety of colored corundum and some pink and red ruby. You will need waders to dig in the creek. The water is deep, 3–4 feet in places, and the current is swift. You can also find pink corundum in the dump piles from the old mine. When you are done digging in the creek, continue up to the old horizontal mine shaft. I would not venture too far into this mine except to take pictures. After you check out the old mine, hike up the mountain about 100 yards above the mine shaft. You will see open pits in the ground where you will find blue/gray corundum pieces on top of the ground after a good rain or you can dig for larger pieces. The corundum mines in this area were opened a short time after the Corundum Hill Mine. They were also mined for gem corundum, and later for industrial corundum.

FEE: There is no fee to collect at this mine, forest service rules apply.

SAFETY: This is a safe place to collect as long as you do not dig in the old mine shaft, the support beams are rotted and dangerous, also keep an eye on small children near the creek.

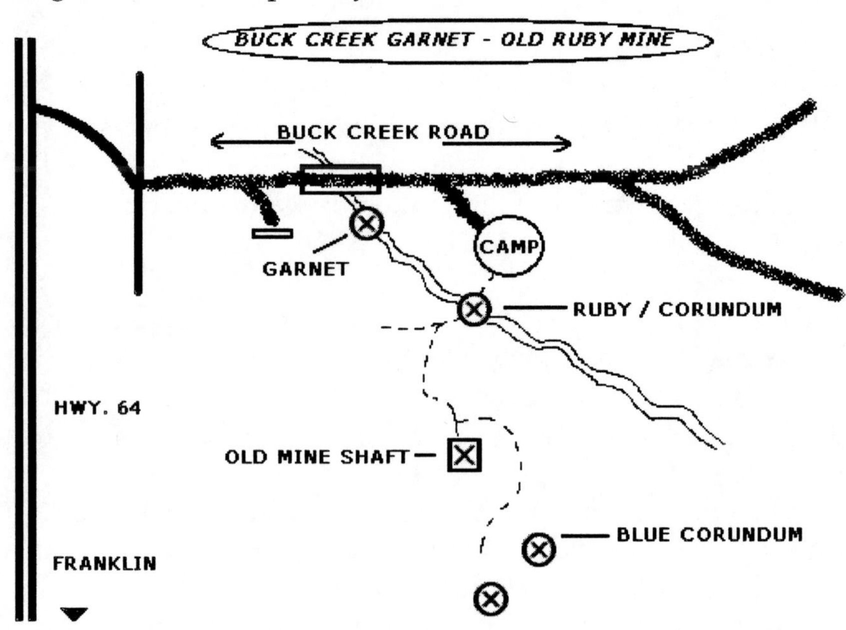

The inside of the old ruby mine on Buck Creek —this is a good photo opportunity but not a safe place to dig

7-pound ruby and corundum specimen from Buck Creek

SITE 21: BUCK CREEK HERBERT CORUNDUM MINE

LOCATION: Clay County, North Carolina.

BEST SEASON: Any, weather permitting.

PROPERTY OWNER: National Forest Service.

MATERIAL TO COLLECT: Hexagonal (6 sided) corundum crystals.

TOOLS: 1/4" sifting screen, shovel, rock pick, rubber boots.

VEHICLE: Any.

DIRECTIONS: From the old ruby mine, continue on Buck Creek Road from the campsite 0.9 miles to the dead end of the road (you will see a sign that says "private drive") park on the left near the locked forest road gate. Hike west/southwest along the old forest road approximately 530 yards. Just over a quarter of a mile, you will come to a field. Look for the trail that cuts through the field to the left of the road, follow the trail approximately 120 yards into the woods. You will see a road that crosses the creek, do not follow it, go right and follow the trail along the creek (Little Buck Creek) southwest approximately 375 yards to the old mine dumps. You will need to cross the creek here to get to the dumps and the old mine shaft.

GPS Coordinates: 35 05.662 N 083 37.527 W (parking area)
GPS Coordinates: 35 05.404 N 083 37.987 W (mine)

WHAT TO LOOK FOR: Dig and sift the material from the dump piles and the creek bed. You will find white and blue corundum pieces and a lot of hexagonal crystals some up to 1"-2" in size. Corundum is plentiful here and you may find some star (asterism) material. This mine was also briefly mined for gem corundum and later for abrasive corundum.

FEE: There is no fee to collect at this mine, forest service rules apply.

SAFETY: This is a safe place to collect, be careful of small children near the old vertical mine shaft, it is full of water and very deep.

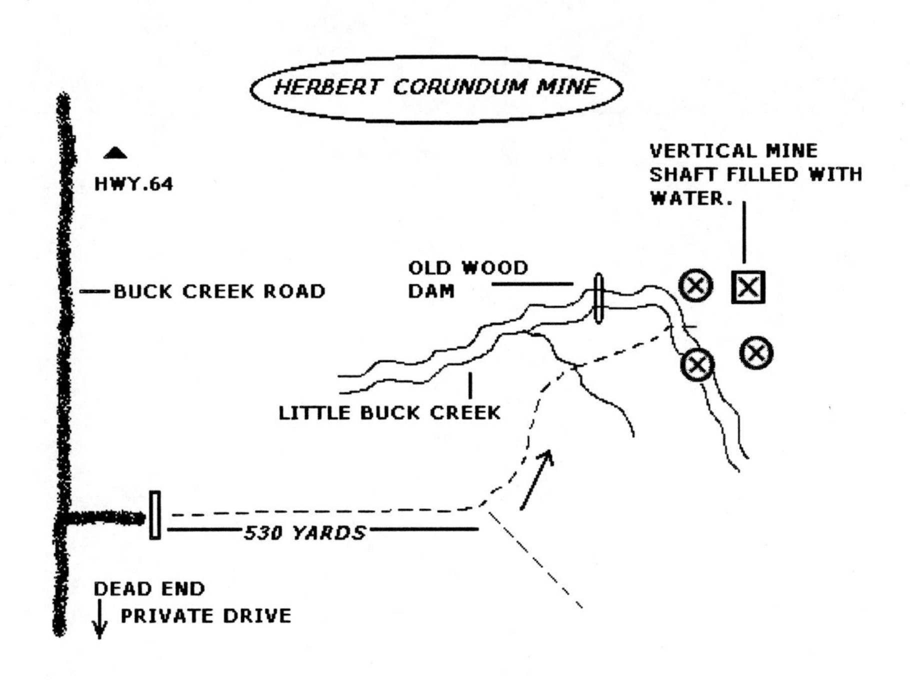

The old Herbert Mine shaft filled with water

SITE 22:
CHUNKY GAL MINE

LOCATION: Clay County, North Carolina.

BEST SEASON: Any, weather permitting.

PROPERTY OWNER: National Forest Service.

MATERIAL TO COLLECT: Ruby in smaragdite, zoisite.

TOOLS: Rock pick, large sledge hammer, 3-lb. sledgehammer, rock chisels, shovel, 1/2" sifting screen.

VEHICLE: Any.

DIRECTIONS: From the intersection of Buck Creek Road and Highway 64 West, continue on 64 West 1.1 miles, turn right onto Old Highway 64 (not marked) make an immediate left onto forest road number 6236, drive 1.0 miles to the parking area at the end of the road.
GPS Coordinates: 35 04.792 N 083 37.656 W (parking area)
GPS Coordinates: 35 04.888 N 083 37.566 W (Chunky Gal)

WHAT TO LOOK FOR: Walk into the woods northeast along the old forest road, you will go about 85 yards and the road will end. Take the path to the right, follow it northeast approximately 200 yards to the top of the hill. (Note: after about 100 yards on this path you will come to an area that has been prospected on your left, you do not want to dig here, continue up the path about another 100 yards and you will come to the main collecting area.) This is a famous collecting site and has been visited by rockhounds from all over the country; there is still plenty of material to collect. The US Govt. has prospect holes and trenches for corundum in numerous areas on the mountain. I have listed the best/most popular places to collect. You will find pink and red ruby in a green/blue colored smaragdite. You can also find gem quality cuttable zoisite. One way to collect is to dig and sift in the dump piles and find nice chunks of pink corundum and sometimes ruby, or you can use your sledge and chisel to crack the boulders that cover the area. There will be veins of pink and red corundum in the rock up to 1/2" wide and sometimes larger. Some 6-sided crystals have been found here as well.

FEE: There is no fee to collect here, forest service rules apply.

SAFETY: This is a safe place to collect, I have come across numerous snakes in the area during the summer and black bears every so often, so keep an eye on children.

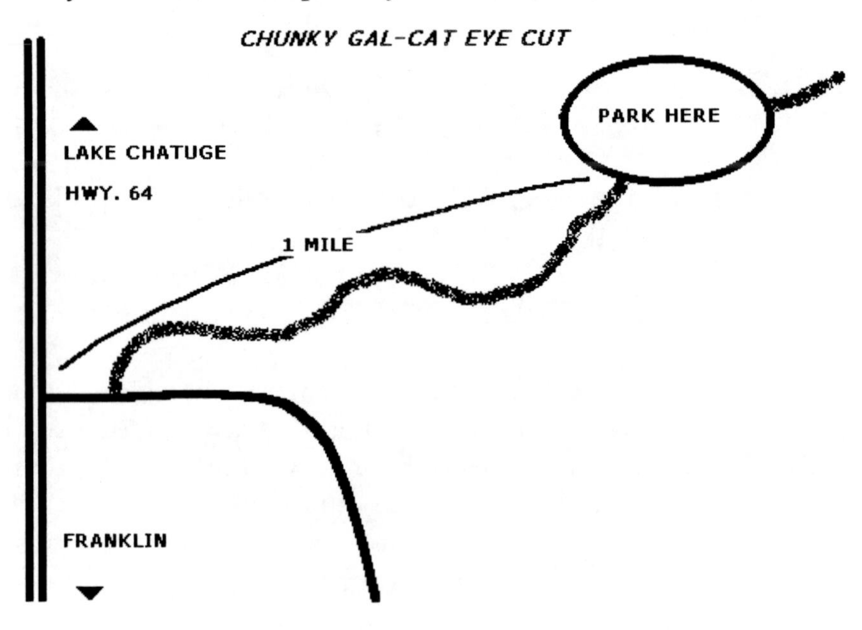

Corundum bearing boulders cover the top of the mountain at Chunky Gal

If you visit Chunky Gal in the summer months, you may find more than mineral specimens!

6-sided ruby crystal in matrix from the Chunky Gal collecting site—crystal is circled

Pink sapphire cabachon set in silver from the Cateye Cut

SITE 23: CAT EYE CUT

LOCATION: Clay County, North Carolina.

BEST SEASON: Fall through spring.

PROPERTY OWNER: National Forest Service.

MATERIAL TO COLLECT: Various colors of corundum/ ruby, common opal.

TOOLS: 1/4" sifting screen, rubber boots, shovel, rock pick.

VEHICLE: Any.

DIRECTIONS: From the Chunky Gal collecting area, return to the old forest road where the path began, take the other path up the hill north over the top of the mountain, you will follow this path down the mountain about 3/4 of a mile, you will see a path that turns to the left through the woods, follow it approximately 100 yards down to the creek at the Cateye Cut.
GPS Coordinates: 35 04.792 N 083 37.656 W (parking area)
GPS Coordinates: 35 05.115 N 083 37.714 W (Cateye Cut)

WHAT TO LOOK FOR: This location is called the "Cat Eye" because much of the corundum found here, when cut will display a star or asterism effect. Here you need to dig in and around the creek bed and screen the material in the creek. The best time to collect here is fall to spring when the water level in the creek is up. In the summer months the creek dries up and it is difficult to identify material. You will find a variety of different colored corundum and some rubies here, most is nice solid cuttable material. You may also find some common opal (yellow and red).

FEE: There is no fee to collect here, forest service rules apply.

SAFETY: This is a safe place to collect, watch for snakes, wild boars and bears.

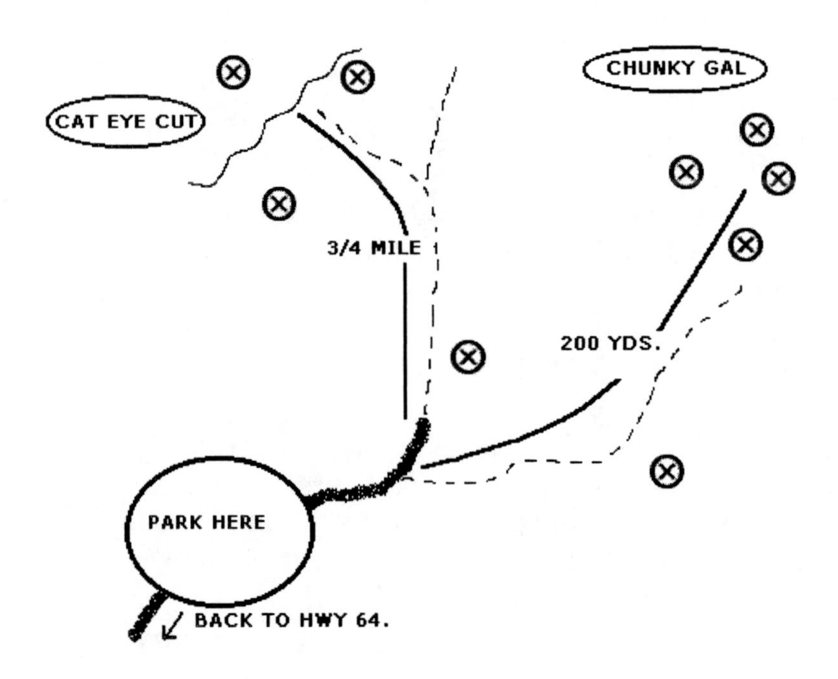

SITE 24: ROCKING CHAIR RUTILE

LOCATION: Clay County, North Carolina.

BEST SEASON: Any, weather permitting.

PROPERTY OWNER: Private.

MATERIAL TO COLLECT: Rutile crystals up to 1".

TOOLS: None.

VEHICLE: Any.

DIRECTIONS: From the intersection of Highway 64 West and Old 64 (the entrance to the Chunky Gal collecting site) continue on 64 West 9.0 miles to Rocking Chair Road, park out of the roadway at the intersection of Rocking Chair Road and 64 West.
GPS Coordinates: 35 01.594 N 083 42.658 W

WHAT TO LOOK FOR: Walk about 50 yards up Rocking Chair Road and into the vacant field on the right. Here, no digging is allowed, but you can surface collect, after a good rain is the best time. You will find nice pieces of rutile up to 1" long, some form nice crystals. I have always received permission to collect here but you can check at one of the houses on Rocking Chair Road to make sure collecting status has not changed.

FEE: There is no fee to collect here.

SAFETY: This is a safe place to collect.

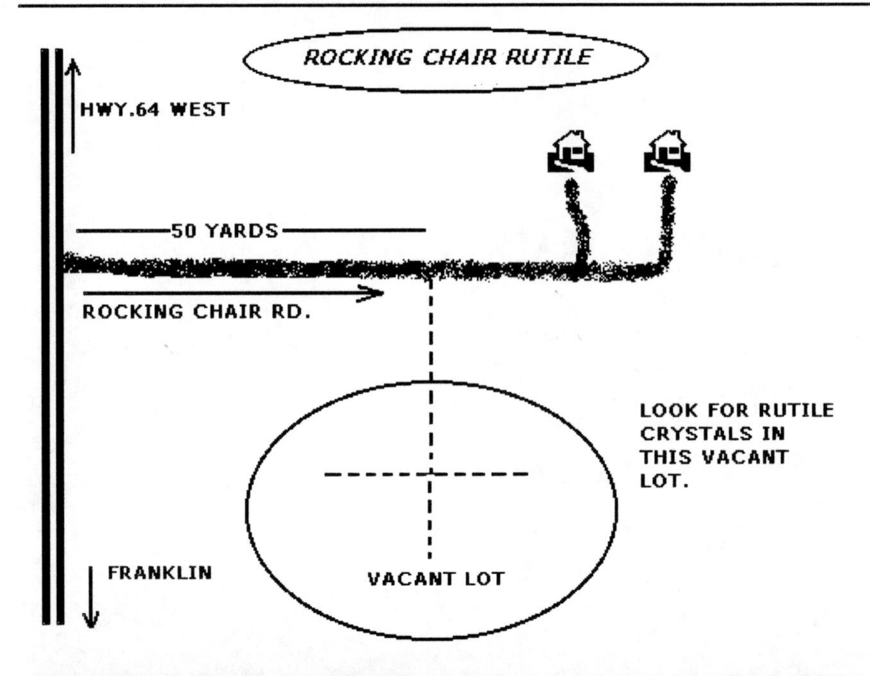

Sections of rutile crystals from the Rocking Chair site

LAKE CHATUGE

Lake Chatuge is 13 miles long and has over 130 miles of shoreline. This is one of my favorite places to collect. I have broken it down into several excellent collecting sites. Some of the minerals and gemstones I have found here include: ruby, sapphire, corundum, apatite, hematite, spinel (pleonast), asbestos, kyanite, limonite pseudomorph cubes, rutilated quartz, tourmaline (schorl), quartz (chalcedony) and quartz crystals. I have also found arrowheads and old pottery pieces. Lake Chatuge is a man made lake which covers an old town. Construction on the dam began in July of 1941 and ended February of 1942. It is maintained by the TVA (Tennessee Valley Authority) and all the lakebed is accessible to the public. The only time to collect here is in the winter months November through February while the lake is at its lowest level. I have also included two sites that are on the Georgia side of the lake.

SITE 25:
LAKE CHATUGE, PENLAND POINT

LOCATION: Clay County, North Carolina.

BEST SEASON: Winter, November through February.

PROPERTY OWNER: TVA.

MATERIAL TO COLLECT: Ruby, sapphire, corundum, apatite crystals in matrix.

TOOLS: Shovel, 1/4" sifting screen, rubber boots.

VEHICLE: Any.

DIRECTIONS: From Franklin, North Carolina, follow 441-23 South, this will turn into Highway 64 West, follow 64 West over the mountain towards the town of Murphy and Hayesville (you will pass the Buck Creek, Chunky Gal, and Rocking Chair collecting sites), as a landmark, from Rocking Chair Road drive 1.8 miles on 64 West to NC Highway 175, turn left and drive 0.3 miles to Elf School Road (SR 1153), turn right and drive 1.0 miles to a fork in the road, stay left, you will come to Penland Point Drive, stay straight onto the gravel road into the campground, follow to Penland Point Lane, turn right, drive 0.1 miles to a parking area on the right (you will see boats and boat trailers parked here), walk southwest about 50 yards onto Penland Point.
GPS Coordinates: 35 01.016 N 083 44.861 W

WHAT TO LOOK FOR: Dig and sift the lakebed sand/ gravel in the lake water to find nice bronze colored sapphires, blue sapphires and rubies. The corundum at this site is very solid, cuttable material. You will also find dark green apatite crystals in a quartz rock matrix. Refer to map on next page for exact collecting locations.

FEE: There is no fee to collect here.

SAFETY: This is a safe place to collect.

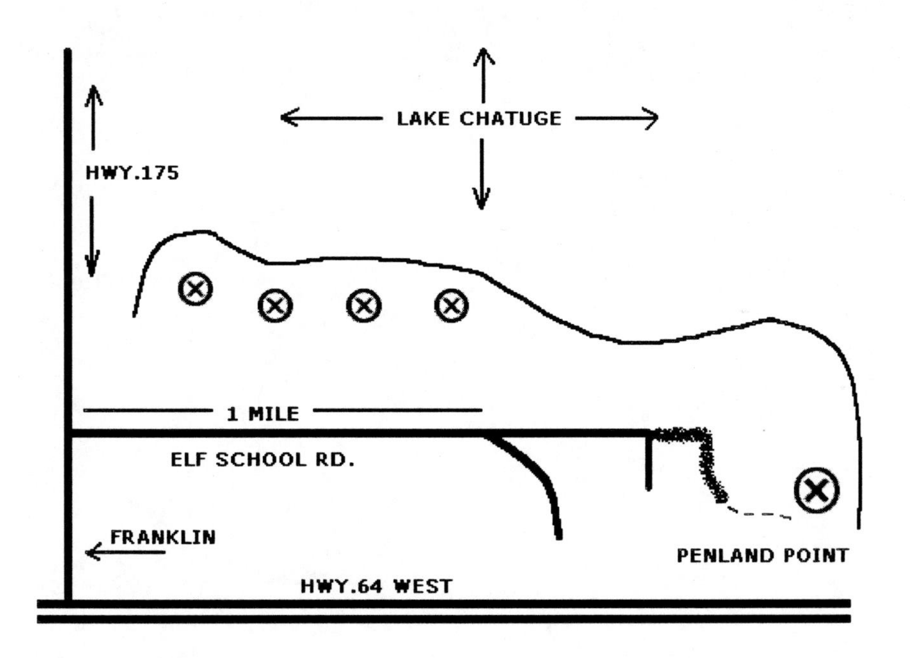

Apatite crystal in matrix from the Penland Point site

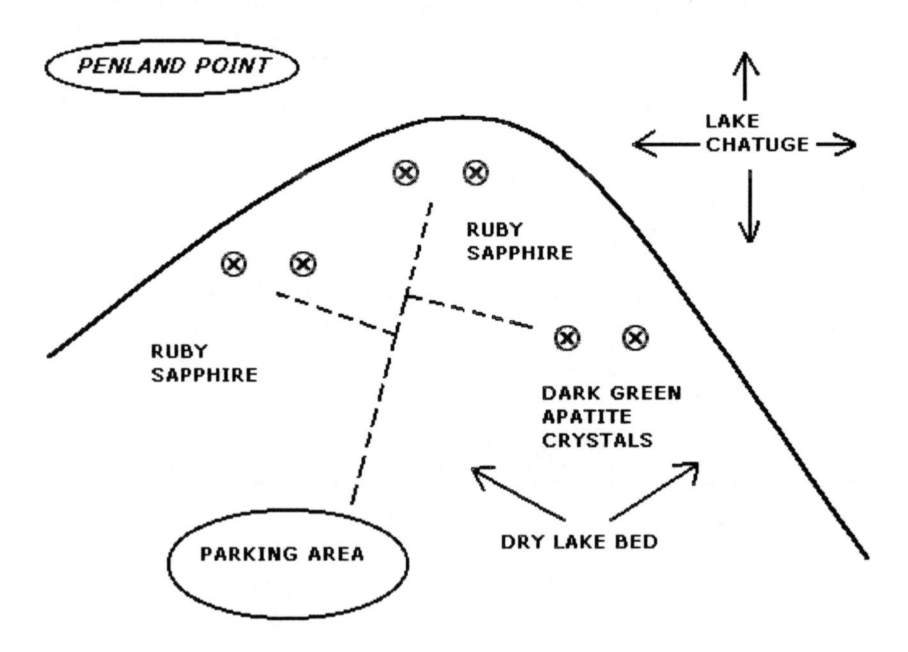

SITE 26:
LAKE CHATUGE, ELF SCHOOL

LOCATION: Clay County, North Carolina.

BEST SEASON: Winter, November through February.

PROPERTY OWNER: TVA.

MATERIAL TO COLLECT: Ruby, sapphire, corundum, quartz crystals, hematite.

VEHICLE: Any.

DIRECTIONS: From the Penland Point collecting site, return to the intersection of Elf School Road and Highway 175, drive 0.3 miles on Elf School Road, park out of the roadway and walk down to the lakebed.
GPS Coordinates 35 01.559 N 083 44.461 W

WHAT TO LOOK FOR: Dig and sift the lakebed sand/ gravel in the lake water to find nice blue sapphires and dark red and pink rubies, also white corundum with blue spots and less frequently clear quartz crystals. Look near the top of the lakebed to find nice specimens of botryoidal hematite. Refer to map for exact locations of each. I have also marked on the map where the old Behr Corundum Mine is located. When the lake is at it lowest point you can see the entrance to the mine under the water.

FEE: There is no fee to collect here.

SAFETY: This is a safe place to collect.

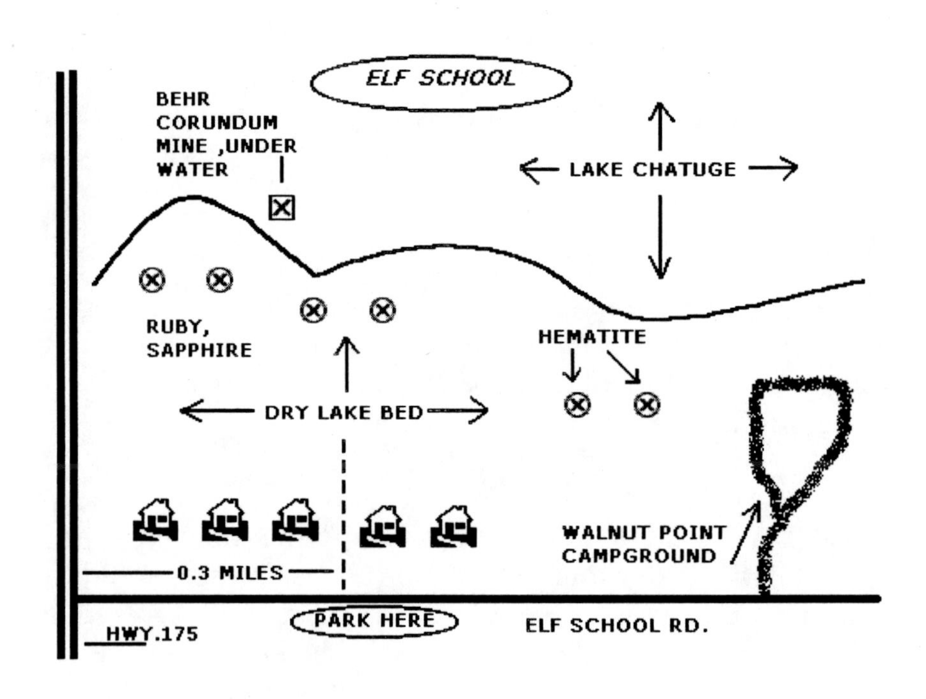

This 3.5-pound corundum specimen was collected from the surface of the lakebed. It is a mass of crystals displaying orange, blue, pink, and white corundum.

Black hematite specimen from the Elf School site

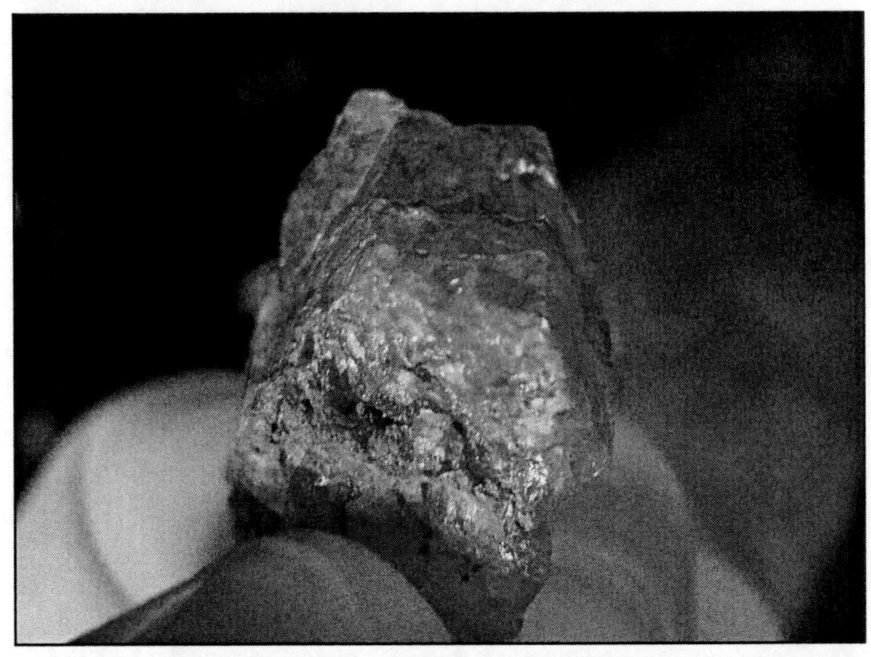

Small quartz crystal from the Elf School site

The lakebed at its lowest water level is the best time to collect

SITE 27:
LAKE CHATUGE, JACKRABBIT RUBY

LOCATION: Clay County, North Carolina.

BEST SEASON: Winter, November through February.

PROPERTY OWNER: TVA.

MATERIAL TO COLLECT: Ruby, corundum.

TOOLS: Shovel, 1/4" sifting screen, rubber boots.

VEHICLE: Any.

DIRECTIONS: From the Elf School collecting site, return to Highway 175, turn right (south) onto Hwy. 175 and drive 0.5 miles, turn right across the bridge over the lake going towards Jackrabbit Campground and the town of Hiawassee, Georgia, this will still be Hwy. 175, drive 2.5 miles to the Jackrabbit Mountain National Forest on the right, turn right, drive 0.4 miles to Philadelphia Lane, turn right and drive 0.3 miles to the end of the road. Get permission from the residence at the end of the road to walk through the cow pasture, walk north-northeast through the field approximately 300 yards to the lakebed.
GPS Coordinates: 35 00.284 N 083 45.629 W

WHAT TO LOOK FOR: Dig and sift the lakebed sand/gravel in the lake water to find pink cube shaped rubies. Look near the top of the lakebed for gray and white corundum and clusters of crystals. Refer to map for exact locations.

FEE: There is no fee to collect here.

SAFETY: This is a safe place to collect.

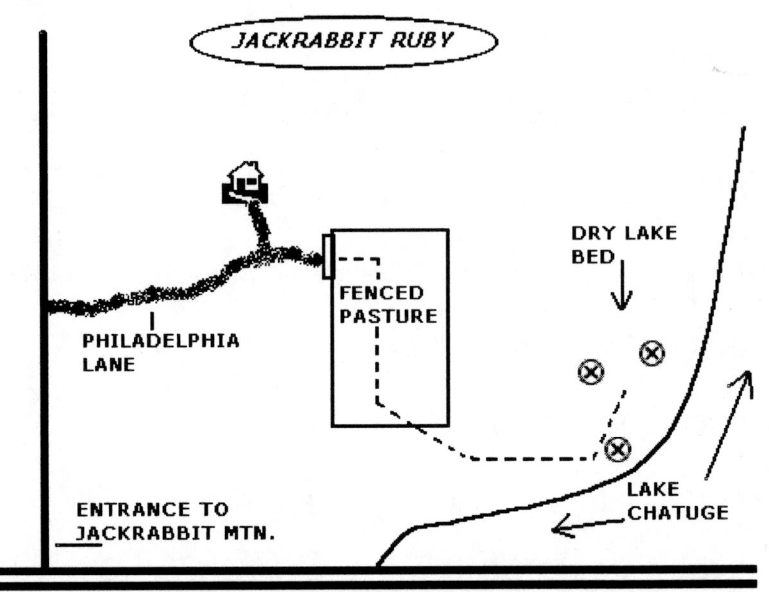

Example of the cube shaped rubies found at the Jackrabbit Ruby site

SITE 28: LAKE CHATUGE, JACKRABBIT LIMONITE CUBES

LOCATION: Clay County, North Carolina.

BEST SEASON: Winter, November through February.

PROPERTY OWNER: TVA.

MATERIAL TO COLLECT: Limonite pseudomorph cubes after pyrite.

TOOLS: Rock pick, shovel, 1/2" sifting screen, rubber boots.

VEHICLE: Any.

DIRECTIONS: From the Jackrabbit Ruby site, return to the beginning of Philadelphia Lane and turn right, continue towards Jackrabbit campground, drive 0.8 miles, just before the boat ramp at the campground turn right into an open field, park on the left of the parking area next to the Jackrabbit Mountain trail sign.
GPS Coordinates: 35 00.677 N 083 46.107 W (parking area)
GPS Coordinates: 35 01.134 N 083 45.986 W (site)

WHAT TO LOOK FOR: Look at the trail signs, you want to follow loop trail (A) (marked in blue) to the 3rd fish attractor access area (these areas are marked by a 4x4 post in the ground), this is approximately 800 yards northeast down the trail. Walk down to the lakebed from the access area and continue northeast approximately another 300 yards. The lakebed will turn to red dirt and clay and you will see a large rock outcropping on the lakebed. Start searching about 200 feet before you get to the rock outcrop for limonite cubes. This is a location I discovered in the winter of 1999. My son, R.J., and I were searching the lakebed for corundum deposits with fellow rockhound, Bill Mintz. I started finding small pieces of dark brown limonite and then some small cubes, then we hit the jackpot. We spent 2-3 hours surface collecting loose cubes and clusters. Some of the cubes were up to 2" on a side. The collecting area covers about 200 feet of the lakebed. I returned to this site two years ago and again found a lot of nice specimens. There is a vein of this material under the lakebed and each summer, when the lake is full, it washes up new material to be collected. You can also dig and sift to find numerous small cubes from 1/2" to 1".

FEE: There is no fee to collect at this site.

SAFETY: This is a safe place to collect.

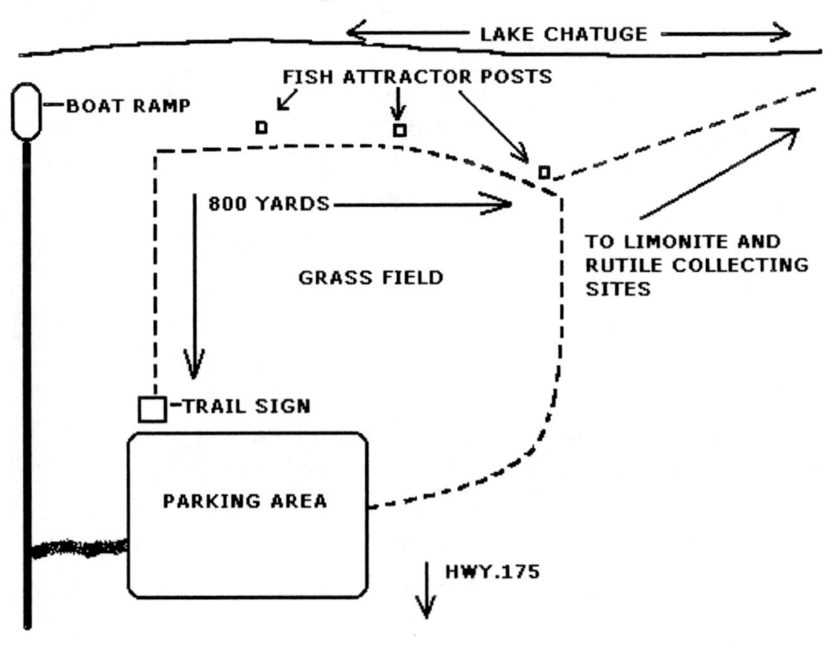

Follow the trail marked on this sign to get to the limonite collecting site

2-inch limonite twin cube from the Jackrabbit site

SITE 29: LAKE CHATUGE, JACKRABBIT RUTILATED QUARTZ

LOCATION: Clay County, North Carolina.

BEST SEASON: Winter, November through February.

PROPERTY OWNER: TVA.

MATERIAL TO COLLECT: Rutilated quartz, tourmaline (schorl), spinel.

TOOLS: Rock pick, 3-lb. sledgehammer, rock chisel.

VEHICLE: Any.

DIRECTIONS: From the Jackrabbit Limonite collecting site, continue walking past the rock outcrop on the lakebed northeast then east approximately 500 yards and begin to look for quartz pieces with rutile and tourmaline.
GPS Coordinates: 35 00.677 N 083 46.107 W (parking area)
GPS Coordinates: 35 01.194 N 083 45.951 W (site)

WHAT TO LOOK FOR: This location was discovered by my son, R.J. We had returned to the Jackrabbit area to look for more limonite and we were still searching for corundum. R.J. found a clear piece of rock and showed it to me, inside were gold colored needles of rutile. We began searching the area and found several pieces of milky quartz with rutile forming on the inside and on the surface of the rock. We also found a chunk about the size of a football that was formed entirely of clear quartz with rutile needles inside. This piece had numerous fractures but we still removed several nice-sized specimens from it. We found small clusters of black tourmaline (schorl). I returned to this site the next winter to collect more rutile and limonite and discovered an unusual specimen—it is 6" in diameter and has a black core with a green matrix chlorite/mica schist around it. I brought the specimen to the US Geological Survey in Asheville, after running several tests it was identified as a massive spinel variety (pleonast). The spinel is opaque black in appearance but when a small piece is chipped off and magnified it has a transparent dark green color. I have searched this area for more of this material and have been unsuccessful; maybe you will get lucky and find the next piece for your collection. The best way to collect here is surface collect, or you can break the larger quartz rock to find rutile inside.

FEE: There is no fee to collect here.

SAFETY: This is a safe place to collect.

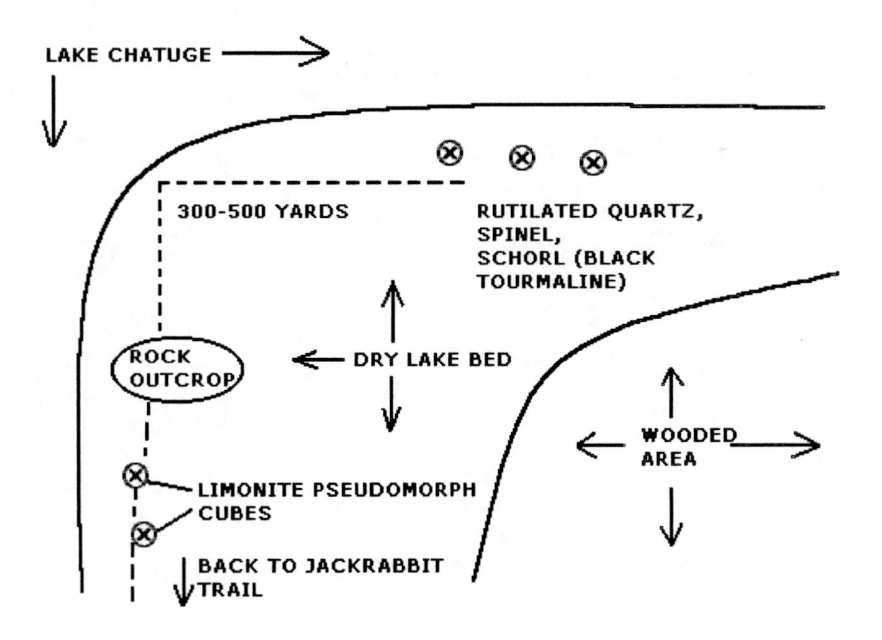

Spinel specimen found on the lakebed at the Jackrabbit rutile collecting site

Rutile needles forming on milky quartz matrix

SITE 30: LAKE CHATUGE, LOWER BELL CHALCEDONY

LOCATION: Towns County, Georgia.

BEST SEASON: Winter, November through February.

PROPERTY OWNER: TVA.

MATERIAL TO COLLECT: Various colors of quartz (chalcedony).

TOOLS: Rock pick, 3-lb. sledgehammer, rock chisel.

VEHICLE: Any.

DIRECTIONS: From the entrance to Jackrabbit Mountain and Highway 175, turn right towards Hiawassee, Ga., drive 1.4 miles to Lower Bell Creek Road, turn right and drive 0.3 miles to the Lower Bell Baptist Church on the right, park in the church parking lot and walk down to the lakebed behind the church.
GPS Coordinates: 34 58.380 N 083 45.079 W

WHAT TO LOOK FOR: This is another location I discovered while searching for corundum. This section of the lakebed is covered with rocks from 1 to 3 feet in size and almost all of them are covered with various colors of chalcedony. If you do not see any color on the outside of the rock, break it open to expose the interior. The chalcedony forms in layers and cavities in the rocks. I have found beautiful cabinet specimens of red, purple, gold and many other colors.

FEE: There is no fee to collect here.

SAFETY: This is a safe place to collect.

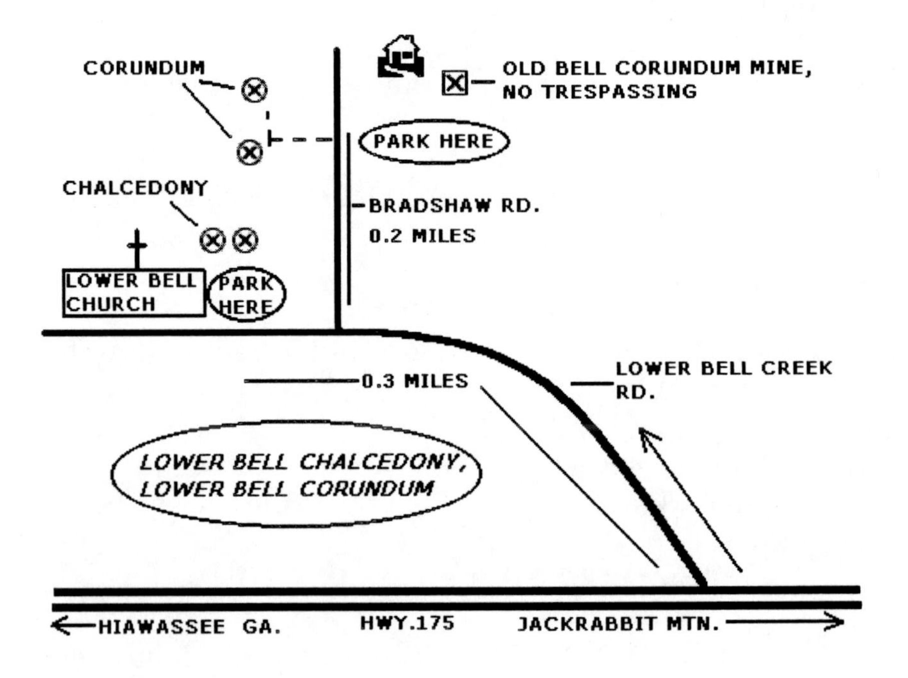

In the winter months, you will find various colors of chalcedony on the lakebed next to the Lower Bell Baptist Church

Quartz, golden chalcedony specimen from the Lower Bell site

SITE 31: LAKE CHATUGE, LOWER BELL CORUNDUM

LOCATION: Towns County, Georgia.

BEST SEASON: Winter, November through February.

PROPERTY OWNER: TVA.

MATERIAL TO COLLECT: Ruby, sapphire, corundum.

TOOLS: Shovel, 1/4" sifting screen, rubber boots.

VEHICLE: Any.

DIRECTIONS: From the entrance to Jackrabbit Mountain and Highway 175, turn right towards Hiawassee, Ga. and drive 1.4 miles to Lower Bell Creek Road, turn right and drive 0.2 miles to Bradshaw Road, turn right and drive 0.2 miles, park out of roadway and walk down to the lakebed. Search 300 yards in either direction on the lakebed from where you park.
GPS Coordinates: 34 58.308 N 083 45.225 W

WHAT TO LOOK FOR: Dig and sift the lakebed sand/gravel in the lake water to find nice red rubies, purple sapphires, and white corundum, refer to map for exact locations. The Old Bell Corundum Mine is located on the hill behind the residence where you park to walk down to the lakebed but no collecting here is permitted.

FEE: There is no fee to collect here.

SAFETY: This is a safe place to collect.

Massive blue corundum from the Lower Bell site

Behind this house at the Lower Bell site is the Bell Corundum Mine

SITE 32: EPWORTH GARNET

LOCATION: Fannin County, Georgia.

BEST SEASON: Any.

PROPERTY OWNER: Private (Willard and Belle Mathis).

MATERIAL TO COLLECT: Giant almandine/iron garnet crystals.

TOOLS: Shovel, rock pick.

VEHICLE: Any.

DIRECTIONS: From Murphy, North Carolina, take US Highway 64-74 West to Highway 60 South, turn left and drive 14.4 miles (you will come to the town of Mineral Bluff, turn left in town and continue on Hwy. 60 South) to US Highway 76-2 West, drive 3.9 miles to the town of Blue Ridge, Georgia, turn right onto Ga. Hwy. 5 North, drive 3.8 miles to Hwy. 2, turn left, drive 5.8 miles to Colwell Road, turn right and drive 1.6 miles to Ritchie Creek Road, turn right onto Ritchie Creek Road and stop at the first house on the left (no. 516), park here.
GPS Coordinates: 34 55.086 N 084 26.702 W

WHAT TO LOOK FOR: Go through the gate to the left of the house (make sure the gate remains shut) and walk approximately 150 yards north into the pasture, you will see a grove of trees in the middle of the pasture, this seems to be the best collecting area. This is the first of three locations in Georgia other than those mentioned at Lake Chatuge that I will include in this book, these sites are only a short drive from Murphy N.C. and I would not want you to miss out on an excellent collecting opportunity while in the Western N.C. area. Here you need to dig in the pasture area, dig in the soil 1–5 feet down to find huge garnet crystals and clusters, some crystals weigh as much as 8–10 pounds, you will also find large weathered staurolite crystals. The owners of the property ask that you fill in all holes to keep the livestock from falling in them. Refer to map for exact collecting location.

FEE: There is a $5.00 per-person per-day fee to collect here. You will need to contact the owner Willard Mathis (423-496-2826) to get permission. Usually the owner will not allow individuals to collect here, you will need to get a large group together and go with them.

SAFETY: This is a safe place to collect.

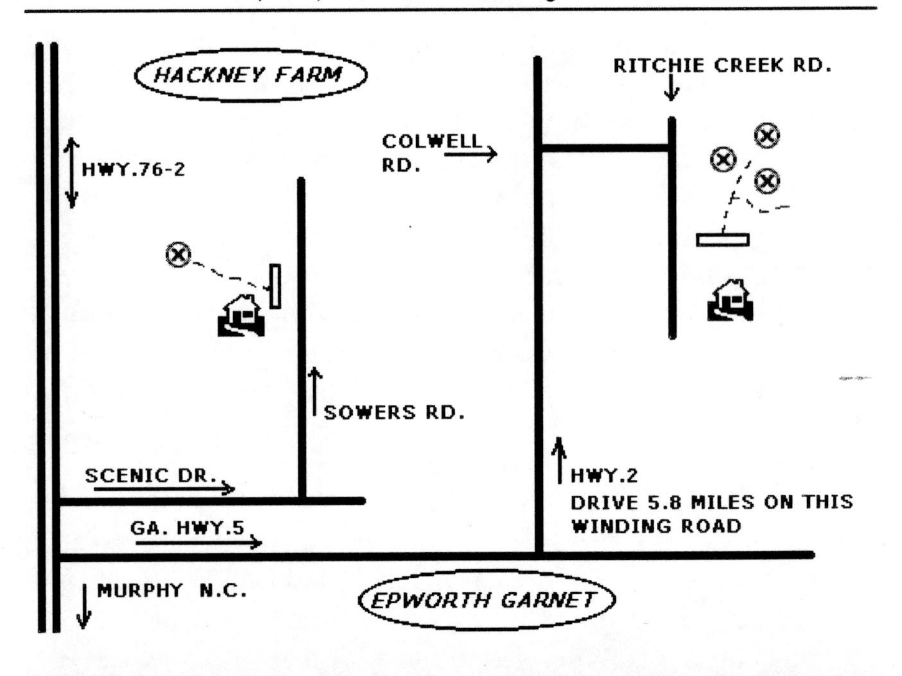

3-pound twin garnet crystal from the Epworth site

SITE 33:
HACKNEY FARM STAUROLITE

LOCATION: Fannin County, Georgia.

BEST SEASON: Any.

PROPERTY OWNER: Private .

MATERIAL TO COLLECT: Staurolite crosses and crystals.

TOOLS: 1/4" sifting screen, shovel, rock pick, bucket.

VEHICLE: Any.

DIRECTIONS: From the Epworth Garnet collecting site, return to the intersection of Highway 5 and Highway 76-2, turn right and drive 0.1 miles to Scenic Drive (CR 217), turn right and drive 1.0 miles to Sowers Road (CR 122), turn left and drive 0.6 miles to the old Hackney residence on the left (no. 611). Just past the house is a gate. If it is open you can drive in to the collecting site. If it is locked you will need to park out of the roadway and walk in to the site. At the time this book was written, the owner of the farm (Steve Hackney) had just recently passed away, the property is being managed by his sister Ann Williams of McCaysville, Ga. There are people renting the house at the farm and I am told that you can leave your collecting payment with them for the owner.
GPS Coordinates: 34 51.989 N 084 20.473 W

WHAT TO LOOK FOR: This second collecting site in Georgia is also a short drive from Murphy N.C. and is probably the most famous staurolite collecting site in Georgia. Dig the soil in the collecting area and fill your buckets, take them to the stream and sift the material to find nice staurolite crosses and single crystal blades—most are small but have perfect shape.

FEE: There is a $5.00 per-person per-day fee for adults and a $2.00 per-person per-day fee for children under 12 years old. Contact Ann Williams at 706-492-5030.

SAFETY: This is a safe place to collect.

SITE 34: COLES CROSSING STAUROLITE

LOCATION: Fannin County, Georgia.

BEST SEASON: Winter, the creekbed is too overgrown in the summer to collect.

PROPERTY OWNER: State of Georgia.

MATERIAL TO COLLECT: Staurolite crosses and single bladed crystals.

TOOLS: 1/4" sifting screen, shovel, rubber boots.

VEHICLE: Any.

DIRECTIONS: From Murphy, North Carolina, take Hwy. 64-74 West to Hwy. 60 South, turn left onto 60 South and drive 10.2 miles to Coles Crossing Road (gravel road), turn right and drive 1.1 miles to the end of the road at the intersection of Coles Crossing and Hardscrabble Road. Park out of the roadway and dig in the creek on the south side of the road.
GPS Coordinates: 34 57.228 N 084 15.303 W

WHAT TO LOOK FOR: This is the last of three locations in this area of Georgia I am including in this book, also a short drive from Murphy N.C. This is an excellent location to find beautiful staurolite crosses and crystals. Park off the road at the intersection of Coles Crossing and Hardscrabble Road. Walk down to the creek, dig in the creekbed and sift the material to find nice well formed staurolite crystals and crosses. Dig on the south side of the road/intersection in the creek. There seems to be more material to be found in this area.

FEE: There is no fee to collect at this site.

SAFETY: This is a safe place to collect.

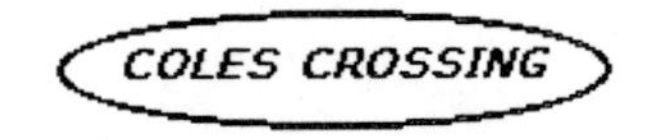

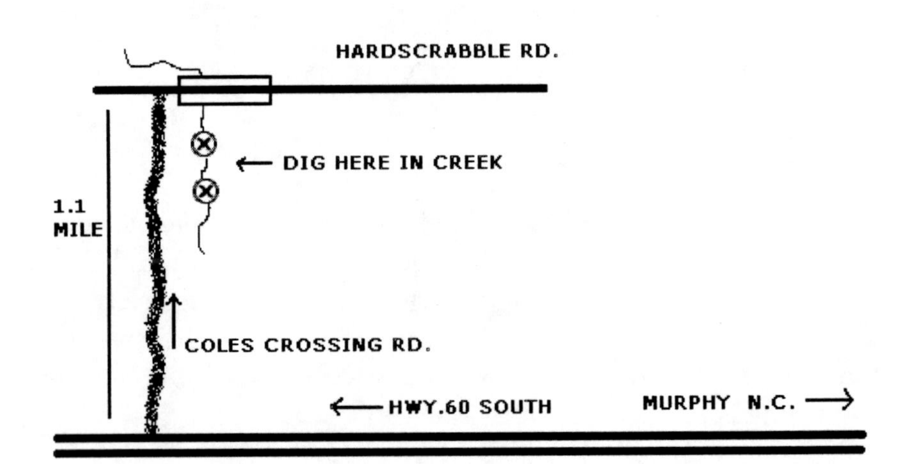

Examples of staurolite crystals from the Coles Crossing site

SITE 35:
COPPER BASIN

LOCATION: Polk County, Tennessee.

BEST SEASON: Any.

PROPERTY OWNER: Tennessee Chemical Company.

MATERIAL TO COLLECT: Pyrite cubes, garnet, actinolite (large crystals), iridescent minerals.

TOOLS: 3-lb. sledgehammer, rock pick, rock chisel.

VEHICLE: Any.

DIRECTIONS: From Murphy, North Carolina, take Highway 64-74 West to the North Carolina/Tennessee line. From the state line drive 2.5 miles into Tennessee, turn right onto Stansbury Mountain Road, drive 0.3 miles to the intersection of Stansbury Mountain Road and Isabella Avenue. Park out of the roadway and walk around the gate into the mine area. You will follow the old road into the mine about 500 yards, where you will come to the old mine buildings. The dump piles and collecting areas are just beyond the old buildings.
GPS Coordinates: 35 01.695 N 084 21.552 W

WHAT TO LOOK FOR: This is a site in Tennessee—just a short drive past the N.C.–Tennessee border from Murphy, N.C. You should take the opportunity to visit here while in Western N.C.

This is an historic mining district that covers several miles and numerous mines. I have given directions to the most popular collecting area. This site was closed for several years by the EPA, but has since been reopened to collecting. Dig anywhere in the numerous dump piles to find pyrite, garnets (almandine), and large specimens of actinolite. I have found some specimens of iridescent hematite. You may also find nickeline and a number of other copper related minerals.

FEE: There is no fee to collect at this site.

SAFETY: This is a safe place to collect, but stay away from the acidic lakes that are in the area.

NOTE: While you are in the area you should visit the historic Burra Burra mine site, it is just a couple of miles past this site on Hwy. 64–74 West.

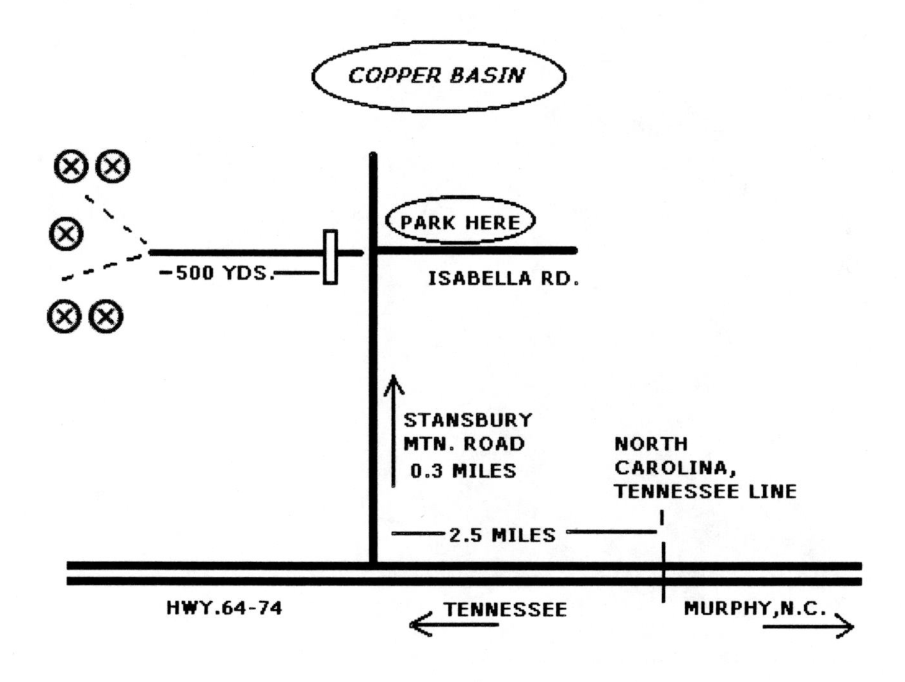

One of the many old mine buildings at the Copper Basin collecting site

Specimen of large actinolite crystals collected at the Copper Basin site

SITE 36:
ALLMON CREEK STAUROLITE

LOCATION: Cherokee County, North Carolina.

BEST SEASON: Any, weather permitting.

PROPERTY OWNER: State of North Carolina.

MATERIAL TO COLLECT: Staurolite crosses and single bladed crystals.

TOOLS: 1/4" sifting screen, shovel, rubber boots.

VEHICLE: Any.

DIRECTIONS: From Murphy, North Carolina take Highway 19-74 towards the towns of Andrews and Marble. When you come to the town of Marble, Highway 141 will turn to the right and Marble Road will turn to the left. Turn left onto Marble Road (SR 1519), drive 0.3 miles to Airport Road (SR 1428), turn right, drive 0.1 miles to Hyatt Creek Road (SR 1379) turn left and drive 0.9 miles to Allmon Creek Road. Turn right onto Allmon Creek Road and cross the bridge, park on the other side of the bridge out of the roadway.
GPS Coordinates: 35 11.407 N 083 55.260 W

WHAT TO LOOK FOR: This area is known for staurolite in the creeks and you could probably find staurolite anywhere on Hyatt Creek or Allmon Creek. I have found the best material at the intersection of these two creeks and a little way up Allmon Creek (about 100 feet). Dig in the creek bed and sift the material to find nice single blades of staurolite and some crosses.

FEE: There is no fee to collect here.

SAFETY: This is a safe place to collect.

Specimens of staurolite from the Allmon Creek site

SITE 37: VENGENCE CREEK STAUROLITE

LOCATION: Cherokee County, North Carolina.

BEST SEASON: Any, weather permitting.

PROPERTY OWNER: State of North Carolina.

MATERIAL TO COLLECT: Large single blades of staurolite, some crosses.

TOOLS: 1/4" sifting screen, shovel, rubber boots.

VEHICLE: Any.

DIRECTIONS: From Murphy, North Carolina take Highway 19-74 towards the towns of Andrews and Marble, N.C. When you get to the town of Marble turn right onto Highway 141, drive 1.0 miles to Vengence Creek Road (SR 1520), park at the intersection of Vengence Creek Road and Hwy. 141 out of the roadway, walk to the south side of the road/bridge and down to the creek.
GPS Coordinates: 35 09.618 N 083 55.209 W

WHAT TO LOOK FOR: Dig in the creek bed and sift the material to find large single blades of staurolite up to 2". You may also find some crosses.

FEE: There is no fee to collect here.

SAFETY: This is a safe place to collect.

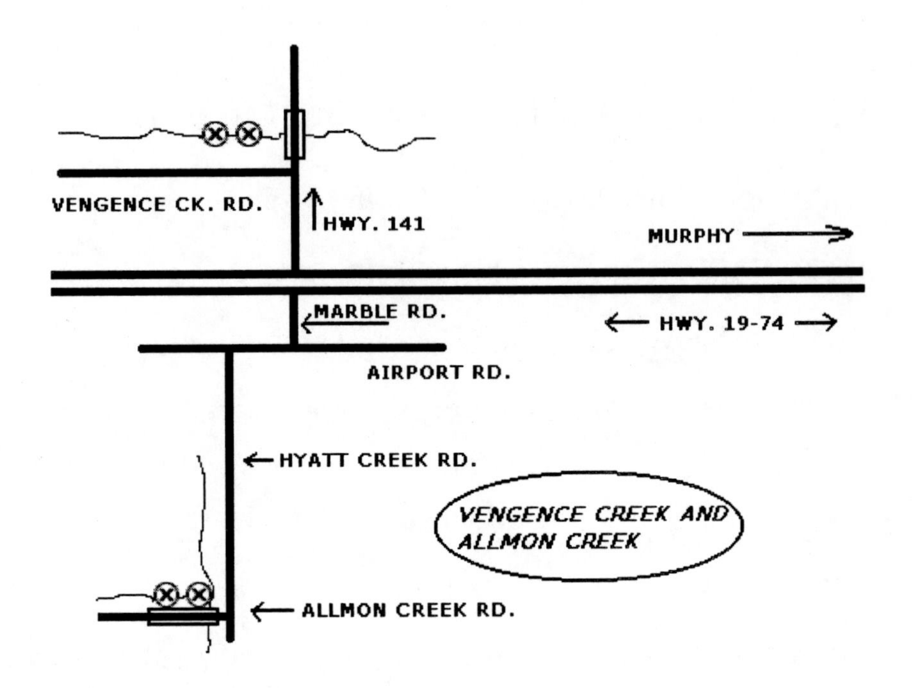
VENGENCE CK. RD.
HWY. 141
MURPHY
MARBLE RD.
HWY. 19-74
AIRPORT RD.
HYATT CREEK RD.
VENGENCE CREEK AND
ALLMON CREEK
ALLMON CREEK RD.

SITE 38:
NANTAHALA TALC & LIMESTONE

LOCATION: Swain County, North Carolina.

BEST SEASON: Any, weather permitting.

PROPERTY OWNER: Private (Nantahala Talc and Limestone).

MATERIAL TO COLLECT: Talc (green soapstone), marble.

TOOLS: 3-lb. sledgehammer, rock pick, rock chisel.

VEHICLE: Any.

DIRECTIONS: From Asheville, North Carolina, take Interstate 40 West to exit 27, Waynesville/Cherokee/19-23-74, follow to exit 103, bear left onto Highway 23-74 Waynesville/Sylva, follow to Bryson City, Highway 23-74 will turn into 64-74 West, follow this highway into the Nantahala gorge. Go through the gorge, turn right onto Hewitts Road (SR 1101), drive 0.5 miles to the quarry entrance.
GPS Coordinates: 35 18.388 N 083 39.081 W

WHAT TO LOOK FOR: This is an active limestone mine that has been mined continuously since it was opened in 1890. From 1895 to 1925 it was mined mainly for talc. Unfortunately, as with many other mines in the Western North Carolina area, it eventually became cheaper to import the talc from a foreign country than to mine it on site. Today the mine produces dolomite limestone gravel, which is sold by the ton. The people who run the mine are very friendly and have always granted me permission to collect. Search the dump piles and you will find nice specimens of talc. You may also find pink, yellow, purple, blue and green-banded marble here. Spend some time and search the whole area.

FEE: There is no fee to collect here. You will need to contact Mr. Jack Herbert, the president of the company (828) 321-3284. An interesting note: Mr. Herbert is only the third president the company has had since it opened 113 years ago.

SAFETY: This is a safe place to collect as long as you stay away from the walls of the mine, watch for falling rock.

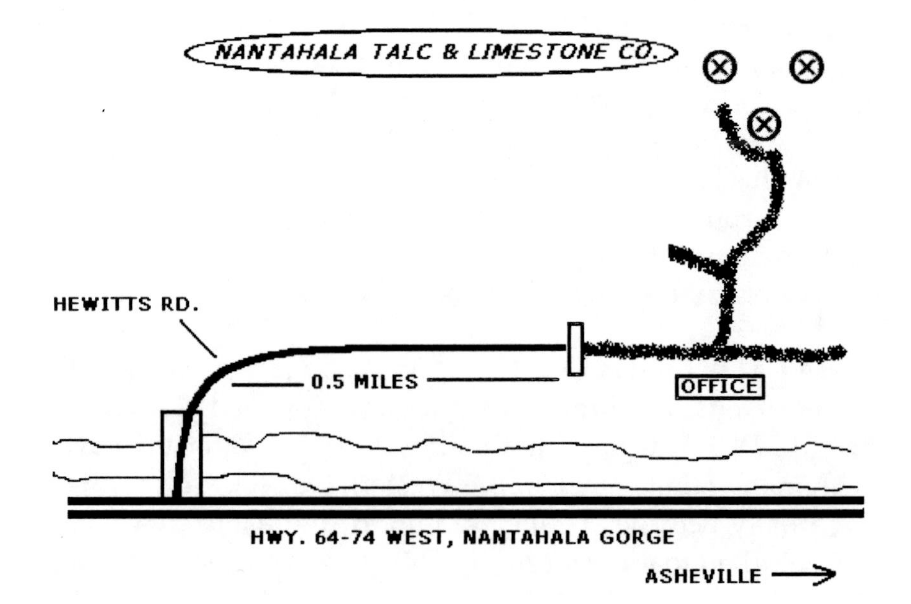

A view of the upper pit at the Nantahala Talc and Limestone Company

Banded pink, yellow, and purple marble from the Nantahala quarry

SITE 39: GRIMSHAWE MINE

LOCATION: Transylvania County, North Carolina.

BEST SEASON: Any, weather permitting.

PROPERTY OWNER: National Forest Service.

MATERIAL TO COLLECT: Ruby, sapphire, corundum.

TOOLS: Rock pick, shovel, 1/4" sifting screen.

VEHICLE: Any.

DIRECTIONS: From Asheville, North Carolina take Interstate 40 West to exit 27 Waynesville/Cherokee/19-23-74, follow to exit 103, bear left onto Highway 23-74 Waynesville/Sylva, take exit 85 Business 23 East Sylva, follow to Highway 107 South, go left, drive 26.7 miles to Highway 64 East, turn left, drive 10.1 miles to NC 281, turn right, drive 4.9 miles to forest road on the right. If you have a four-wheel drive vehicle you can follow the road to the right to the collecting area, if not, park and hike the forest road approximately 300 yards to the mine.
GPS Coordinates: 35 04.207 N 083 00.559 W

WHAT TO LOOK FOR: You will come to a ravine/cut in the road, if driving, park here. You can continue up the road on foot and dig and dry sift in the dump piles by the old mine diggings on the right or follow the trail into the woods to the left of the ravine. You will see prospecting holes throughout the area. You need to dig to the gravel layer near or in the creek bed and sift the material in the creek. You will find sapphires in a variety of colors and nice rubies. You can also find nice gem almandine garnet. Some of the corundum here is facet-grade material. The Grimshawe was originally mined for asbestos in the 50s and 60s.

FEE: There is no fee to collect at this mine, forest service rules apply.

SAFETY: This is a safe place to collect.

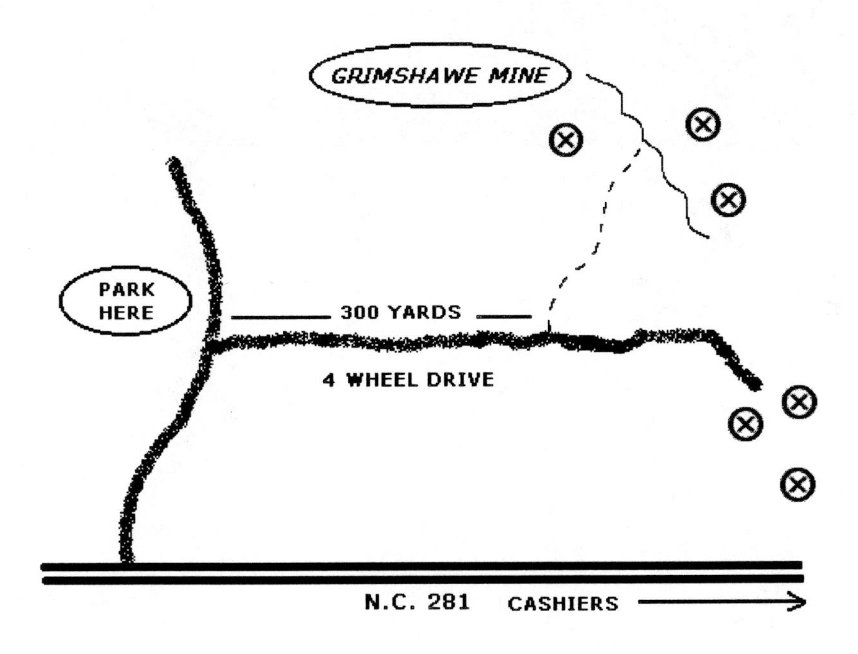

Some facet-grade sapphires from the Grimshawe collecting site (the colors are green, deep blue, pink and yellow)

SITE 40: SHEEPCLIFF AQUAMARINE MINE

LOCATION: Jackson County, North Carolina.

BEST SEASON: Any, weather permitting.

PROPERTY OWNER: Private.

MATERIAL TO COLLECT: Beryl crystals, aquamarine, smoky quartz, yellow, orange, pink feldspar.

TOOLS: Rock pick, 3-lb. sledgehammer, rock chisel, 1/2" sifting screen.

VEHICLE: Any.

DIRECTIONS: From Asheville, North Carolina, take Interstate 40 West to exit 27 Waynesville/Cherokee/19-23-74, follow to exit 103, bear left onto Highway 23-74 Waynesville/Sylva, take exit 85 Business 23 East Sylva, follow to Highway 107 South, turn left, drive 22.3 miles to Bee Tree Road (SR 1124), (note: there are two left turns for Bee Tree Road, you will pass the first turn for Bee Tree Road, do not turn here, take the second turn for Bee Tree Road), turn left, drive 0.4 miles to Ceder Creek Road (SR 1120), turn right, drive 3.1 miles to the Treasurewood Subdivision (you will come to a fork in the road at Shirley Pressley Road, stay right at the fork, the road will turn to gravel, follow it to Treasurewood), turn right into Treasurewood, drive 0.6 miles (you will come to a fork in the road, stay to the left onto the gravel road) to the entrance gate, drive 0.3 miles to lot number 8 on the left, park on the

right in the fenced parking area. If the gate is closed you will have to walk the 0.3 miles to the mine.
GPS Coordinates: 35 80.153 N 083 05.453 W

WHAT TO LOOK FOR: After you park, walk across the road into the woods on the trail. You will follow the trail south-southeast approximately 100 yards where you will come to the dump piles of the mine. This was an active aquamarine mine years ago and today you can still find beautiful specimens of beryl and aquamarine. I have also found nice specimens of clear smoky quartz crystals and feldspar in various colors. This property was bought a few years ago by a residential developer but at this time collecting is still allowed. You can dig and sift the dump pile material to find the smaller beryl crystals or break the larger rocks to find specimens in matrix.

FEE: There is no fee to collect at this site.

SAFETY: This is a safe place to collect. Watch small children near the old vertical mine–it is full of water and deep.

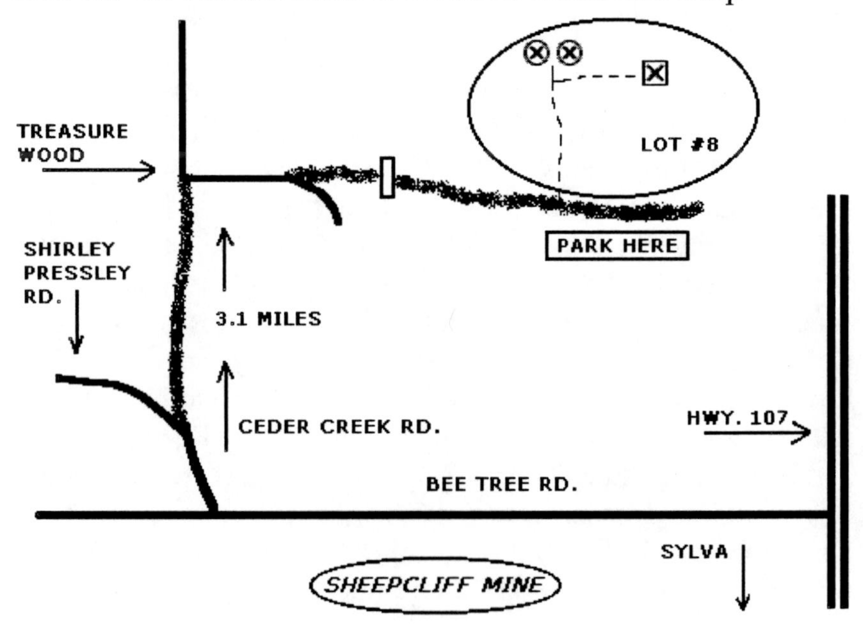

The Sheepcliff Aquamarine Mine is today filled with water

Beryl specimen in matrix from the Sheepcliff mine

Dig through the dump piles at the Sheepcliff Mine to find nice specimens of blue aquamarine

Quartz crystals from the Sheepcliff Mine

SITE 41: BLACK MOUNTAIN KYANITE & SAPPHIRE

LOCATION: Buncombe County, North Carolina.

BEST SEASON: Any, weather permitting.

PROPERTY OWNER: Private.

MATERIAL TO COLLECT: Small to large size boulders/ masses of kyanite with sapphire.

TOOLS: 3-lb. sledgehammer, rock chisels, rock pick.

VEHICLE: Any.

DIRECTIONS: From Asheville, North Carolina, take Interstate 40 East to exit 64, turn left onto NC Highway 9 North, follow this into town, turn right onto US Highway 70 East, follow to Flat Creek Road (SR 2515) turn left and immediately right onto E State Street-US 70, drive 0.5 miles to McCoy Cove Road, turn left, drive 0.8 miles to Charmeldee-Sky Hi Acres residential neighborhood, park out of the roadway near the entrance to Charmeldee.
GPS Coordinates: 35 37.832 N 082 18.105 W

WHAT TO LOOK FOR: Walk east down the hill into the woods past the houses on McCoy Cove Road. Behind the houses about 75 yards is a small creek. Follow the creek bed south and look for masses of kyanite. The kyanite is a pale blue color and some contains nice blue sapphires up to 1/4" in size. I have seen some sapphires from here that were 1" across but these are rare. Look closely at the rocks in the creek, some of the kyanite is black from weathering but it will clean up. You can find this material approximately 1000 yards south along the creek from where you park. The residents along the road here have always granted me permission to collect, you may ask at one of the houses along the road to double check. You will need to break the larger boulders down into manageable size pieces to remove them.

FEE: There is no fee to collect here.

SAFETY: This is a safe place to collect.

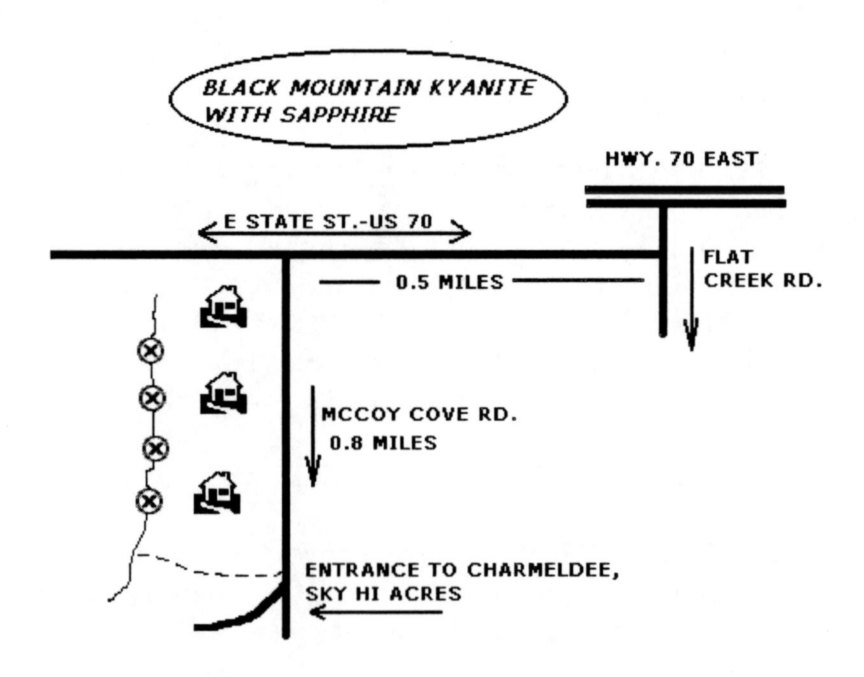

116-pound boulder of pale blue kyanite
with dark blue sapphire crystals
collected in the creek at the Black Mountain site

SITE 42: RIDGECREST KYANITE

LOCATION: Buncombe County, North Carolina.

BEST SEASON: Any, weather permitting.

PROPERTY OWNER: Norfolk Southern Railway.

MATERIAL TO COLLECT: Pale blue kyanite in blades and as masses.

TOOLS: Rock pick.

VEHICLE: Any.

DIRECTIONS: From Asheville, North Carolina, take Interstate 40 East to exit 66 (Ridgecrest), turn right at the stop sign, cross the bridge, turn right onto Old US 70 East Extension, drive 0.8 miles to a railroad crossing, park in the gravel area to the left of the road (north side) of the train tracks.
GPS Coordinates: 35 37.090 N 082 18.031 W

WHAT TO LOOK FOR: Walk east along the tracks and search the cuts on both sides of the tracks. You will find nice masses of pale blue kyanite and some single pieces.

FEE: There is no fee to collect here.

SAFETY: This is a safe place to collect for adults, keep an eye out for trains. I would not bring children here.

NOTE: While in the downtown Asheville area, I would suggest you visit the Colburn Gem and Mineral Museum located at Pack Place. They have a nice collection of local gems and minerals including the Star of the Carolinas from the Old Pressley Sapphire Mine. If you are interested in purchasing gem and mineral specimens while in the area, I would suggest the Silver Armadillo in the Westgate shopping mall in West Asheville. This is the largest rock, gem and mineral store in the area.

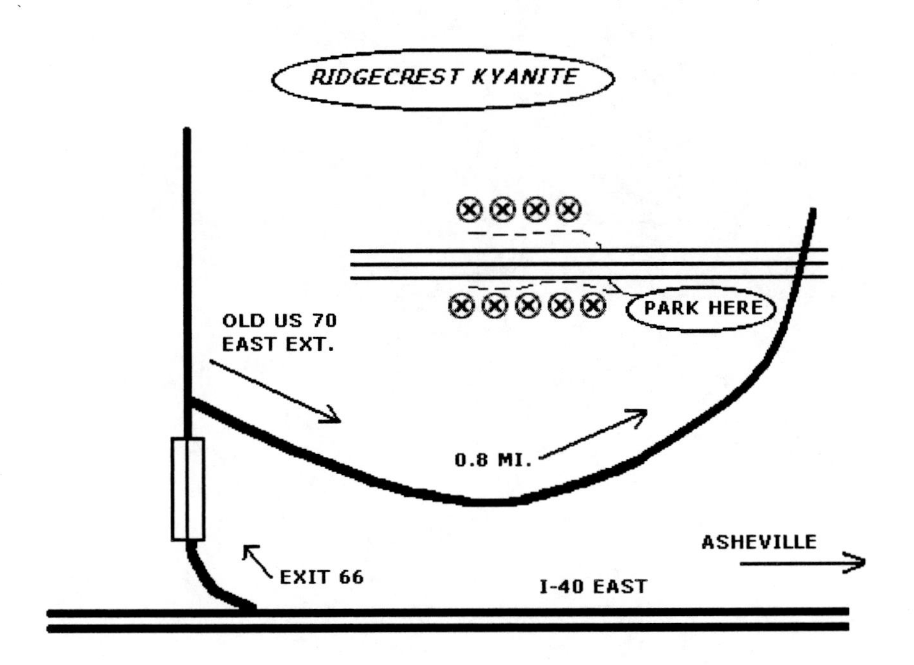

Search both sides of this railroad cut to find masses of pale blue kyanite. Be careful of the trains!

SITE 43: WOODLAWN QUARTZ CRYSTALS

LOCATION: McDowell County, North Carolina.

BEST SEASON: Any, weather permitting.

PROPERTY OWNER: Private (Woodlawn Quarry).

MATERIAL TO COLLECT: Clear quartz crystals, phantom quartz crystals.

TOOLS: Shovel, 1/2" sifting screen, rock pick.

VEHICLE: Any.

DIRECTIONS: From Asheville, North Carolina, take Interstate 40 East to exit 85, take NC Highway 226 North, this will turn into NC 221 North, from exit 85 drive 12.3 miles, turn right onto American Thread Road (SR 1556) you will see a sign that says (Coats North America) at the turn, drive 0.2 miles to the entrance to the quarry on the right.
GPS Coordinates: 35 47.233 N 082 02.167 W

WHAT TO LOOK FOR: This is an active limestone quarry. Once inside, proceed to the office to get permission to collect, they will direct you where to park. The collecting area is not in the quarry itself but in the woods on the hill behind the quarry. Follow the trails through the woods and you will see numerous holes made by previous rockhounds. This area is a weathered limestone. Dig in the soft dirt and screen the material to find perfect clear quartz crystals. You may also find some nice clusters but the crystals tend to be loose so be careful when wrapping your specimens.

FEE: There is no fee to collect at this site.

SAFETY: This is a safe place to collect.

NOTE: If you have the time after you leave the Woodlawn Quarry you can return to Hwy. 221 North and follow the signs to Grandfather Mountain. This is a great place to visit with a small zoo and museum and they have one of the best collections of North Carolina minerals in the area. I highly recommend the mineral museum.

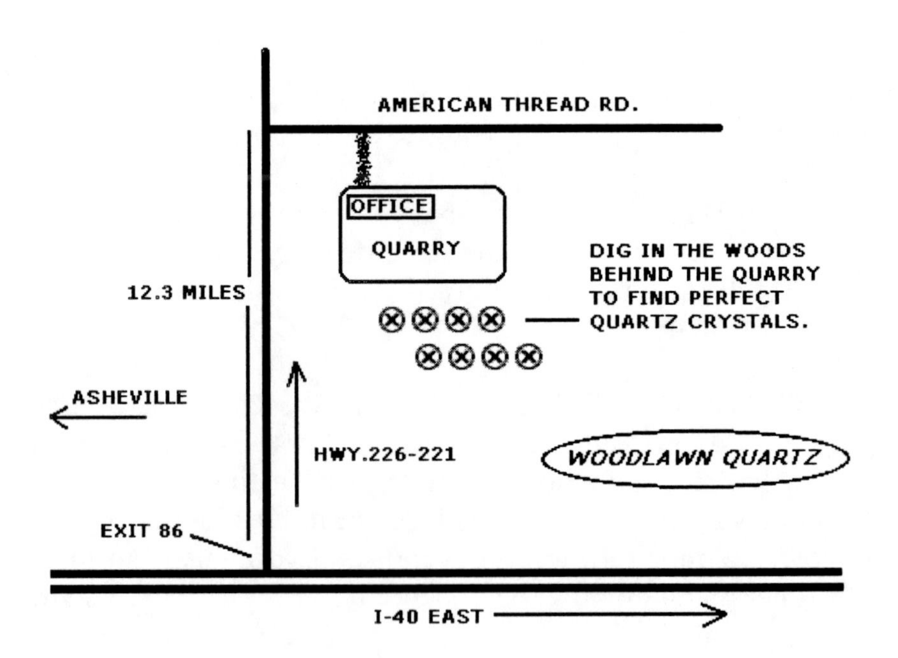

In the woods behind the Woodlawn Quarry is where you will find numerous perfect quartz crystals

Quartz crystal specimen from the Woodlawn Quarry

SITE 44: EMERALD HOLLOW MINE

LOCATION: Alexander County, North Carolina.

BEST SEASON: Open year round 8:30 a.m. until sunset.

PROPERTY OWNER: Private (Mike and Dotty Watkins).

MATERIAL TO COLLECT: Emerald crystals, hiddenite, smoky quartz, tourmaline (schorl), aquamarine, sapphire.

TOOLS: You may rent tools at the site or bring your own, I would suggest: shovel, rock pick, 3-lb. sledgehammer, rock chisels.

VEHICLE: Any.

DIRECTIONS: From Asheville, North Carolina, take Interstate 40 East to exit 148 (US Highway 64), follow Highway 64, 12 miles into the town of Hiddenite, turn right onto Sulpher Springs Road (SR 1001), follow the green mine signs to the mine on the right.

WHAT TO LOOK FOR: This mine is a bit to the east of Western North Carolina, but is included because of the recent emerald finds in this area. Millions of dollars' worth of emeralds are being found by a local man, James Hill, from Alexander County at the old Rist Ellis Mine. At the Emerald Hollow Mine, you can dig in the same pegmatite system. This mine is salted if you buy the buckets of dirt they sell, but you can dig elsewhere on the property and find some nice native material, the minerals I listed above are native to this area. Another nice thing about this mine is, if you locate a possible mineral/gem bearing vein the owners will let you stake a claim for a few dollars a day, a few days working a good vein and you might have your own million dollar emerald!

FEE: Varied

SAFETY: This is a safe place to collect.

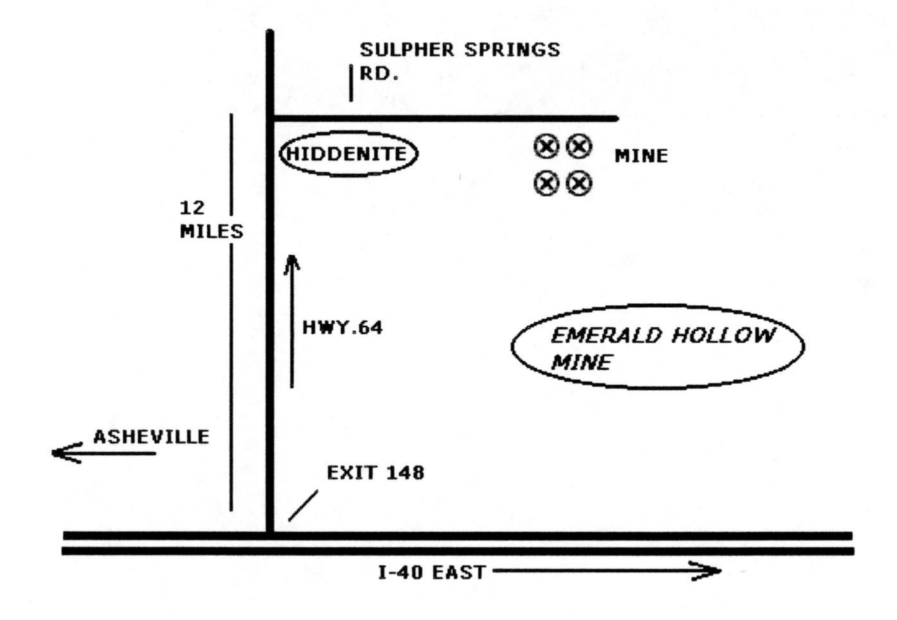

These emerald crystals are displayed in the Smithsonian Museum in Washington, DC. They were collected in Hiddenite, NC

SITE 45: POOVEY GARNET MINE

LOCATION: Burke County, North Carolina.

BEST SEASON: Any, weather permitting.

PROPERTY OWNER: Private (Tiffany Poovey).

MATERIAL TO COLLECT: Large 12- and 24-sided almandine garnet crystals.

TOOLS: Rock pick, 3-lb. sledgehammer, rock chisel, shovel.

VEHICLE: Any.

DIRECTIONS: From Asheville, North Carolina, take Interstate 40 East to exit 105 (Morganton/Shelby) turn left at stop light onto NC 18 South, drive 2.9 miles, turn right onto Port Street (SR 1929), drive 0.1 mile to the Poovey residence on the left, number 3290. Pay the owner here, if no one is home leave payment in the screen door. After you pay, return to Hwy 18 South, turn right and drive 2.6 miles, park on the side of the road. Pay close attention to your odometer to find this site, when you park you will not see anything that looks like a mine, the digging area is on the east side of the road about 20 feet up a small hill, when you get to the top of the hill you will see the holes left by previous rockhounds.
GPS Coordinates: 35 40.740 N 081 34.455 W

WHAT TO LOOK FOR: This mine had been closed since the late 90s, when the owner, Mr. Poovey, died. Only recently did his daughter, Tiffany, begin to allow collecting here again. You will need to dig down 3–5 feet in the ground to find the larger garnet crystals, some of the crystals are gem quality. You may find 12- and 24-sided specimens, some as big as softballs. If you want matrix specimens dig in the sides of the pit to find crystal clusters in the schist rock.

FEE: There is a $10.00 per-bucket, per-person fee to collect here, which is very reasonable, it will take you a while to fill a bucket.

SAFETY: This is a safe place to collect.

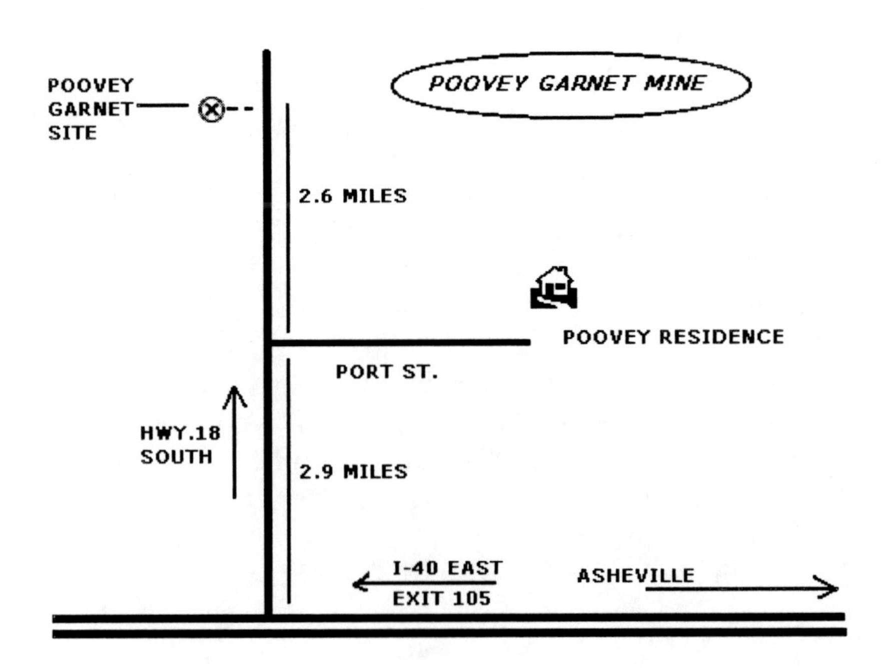

24-sided specimen of an almandine garnet collected at the Poovey site

SITE 46:
RAT TAIL CORUNDUM

LOCATION: Lincoln County, North Carolina.

BEST SEASON: Any, weather permitting.

PROPERTY OWNER: Private.

MATERIAL TO COLLECT: Bronze, gray, corundum covered in sillimanite.

TOOLS: Rock pick, shovel, 1/2" screen.

VEHICLE: Any.

DIRECTIONS: From the Poovey Garnet Mine continue on Hwy. 18 South, drive 10.8 miles, turn right onto CR 1103 towards South Mountain State Park, turn right at the stop sign onto CR 1100, drive 0.7 miles, turn left onto Allran Road (gravel road), drive 0.4 miles to a private drive on the right (317) follow driveway through the woods to owners house on the left, get permission to collect here, park and walk south down the road to an old wooden shed turn right and walk west following the power line cut about 100 yards, you will see the collecting area on the right.
GPS Coordinates: 35 33.448 N 081 31.527 W

WHAT TO LOOK FOR: As far as I know this is the only place to find corundum in this form. It is called rat tail because the sillimanite coating is gray and most specimens tend to be pointed on the ends resembling a rat's tail. You may also find specimens with the end of the corundum crystal inside showing. You will need to dig down in the soft dirt 1 to 2 feet to find this material. I would sift with a 1/2" screen to make sure you get the smaller crystals.

FEE: There is no fee to collect here.

SAFETY: This is a safe place to collect.

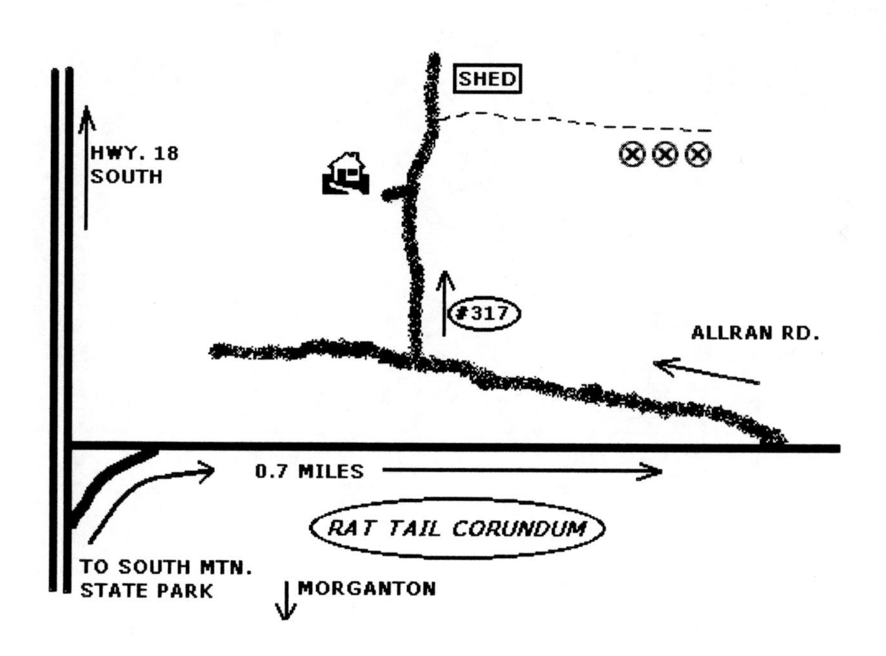

Corundum crystals covered in sillimanite, resembling a rat's tail. Also a 2-inch hex crystal.

SITE 47:
LINCOLN COUNTY SMOKY QUARTZ

LOCATION: Lincoln County, North Carolina.

BEST SEASON: Any, weather permitting.

PROPERTY OWNER: Private.

MATERIAL TO COLLECT: Smoky quartz crystals.

TOOLS: Rock pick.

VEHICLE: Any.

DIRECTIONS: From the Rat Tail Corundum collecting site return to Hwy. 18 South, turn right, drive 0.7 miles to NC Hwy. 10 East, turn left, drive 4.3 miles to Rhoney Farm Road, turn left and immediately right into the vacant lot (lot is made of red clay dirt).
GPS Coordinates: 35 35.045 N 081 26.382 W

WHAT TO LOOK FOR: I have collected here for about three years. This site was discovered by a local rockhound and friend (Bill Mintz). I collect on the east side of the lot on the red clay/dirt bank, the bank is about ten feet high and you will see a quartz vein running through the dirt. Dig in the vein and you will find small 1–2" smoky quartz crystals. The crystals are a yellow-brown color.

FEE: There is no fee to collect here.

SAFETY: This is a safe place to collect.

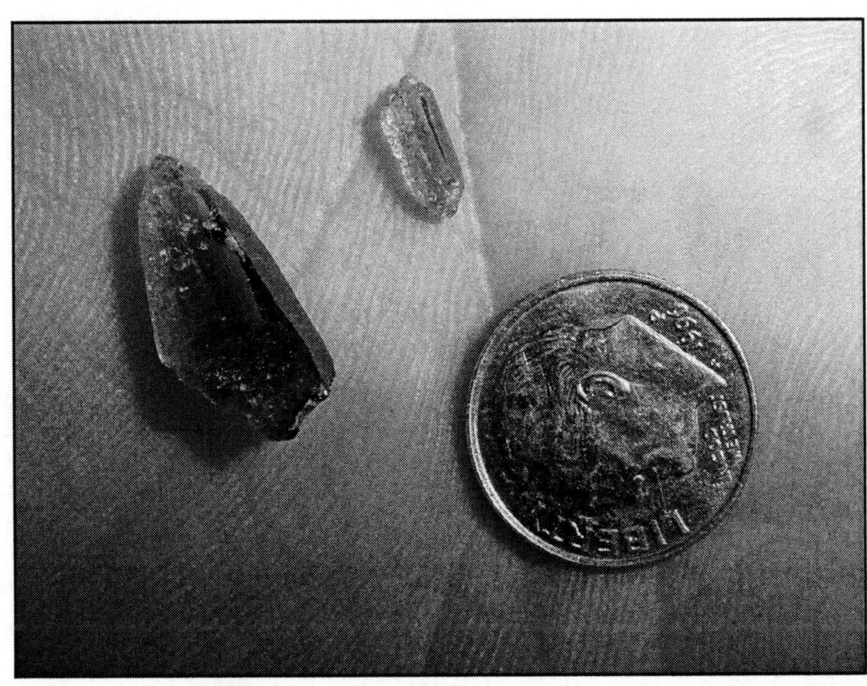

Specimens of quartz crystal from the Lincoln County site

SITE 48:
PROPST CORUNDUM

LOCATION: Lincoln County, North Carolina.

BEST SEASON: Any, weather permitting: I would suggest wintertime.

PROPERTY OWNER: Private (Propst family).

MATERIAL TO COLLECT: Ruby and sapphire hexagonal crystals and masses.

TOOLS: Rock pick, prybar, 3-lb. sledgehammer, rock chisel, shovel, 1/4" sifting screen.

VEHICLE: Any.

DIRECTIONS: From the Lincoln County Quartz location on Hwy. 10 East, continue on Hwy. 10 East, drive 12.5 miles to Startown Road, from the intersection of Startown Road and Hwy. 10 East, turn right, drive south on Startown Road for 6.6 miles, you will see a brick house on the left side of the road (3696 Propst), park in the driveway and go into the open carport, you will see a tin can next to the back door, put your collecting money in the can, drive to the back of the house behind the old metal building in the field and park here, walk into the field behind the metal building and you will see the holes from the previous rockhounds, pick a spot and start digging.
GPS Coordinates: 35 33.275 N 081 16.167 W

WHAT TO LOOK FOR: This is one of my favorite places to collect rubies and sapphires. I have seen 6-sided crystal specimens from this location 6" long and 4" across. I have also seen some nice crystal clusters and have found hundreds of nice hex crystals for my collection. If you come here after a big rain, you can dig and sift for smaller crystals in the holes that will be full of water. The best time to collect is in the winter when it is dry. You will need to dig down to the quartz gravel layer and search for the corundum crystals. Bring an extra set of clothes to change into when you are done digging. This area is made of red clay dirt and you will get very dirty but it will be worth it.

FEE: There is a $5.00 per-person per-day fee to collect here.

SAFETY: This is a safe place to collect.

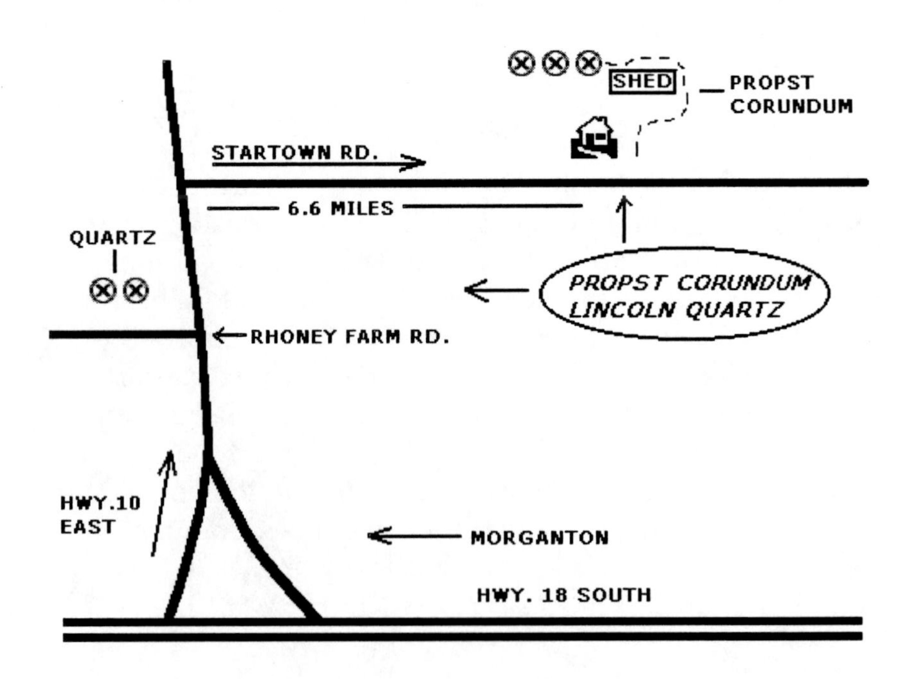

Dig just beyond this old metal shed in the field to find nice ruby and sapphire specimens

2.5" pink and purple sapphire crystal from the Propst site

SITE 49: YATES BROOKS FARM

LOCATION: Cleveland County, North Carolina.

BEST SEASON: Any, weather permitting.

PROPERTY OWNER: Private (Whittaker family).

MATERIAL TO COLLECT: Muscovite mica crystals, anatase, rutile, quartz crystals.

TOOLS: Shovel, 1/2" sifting screen, rock pick.

VEHICLE: Any.

DIRECTIONS: From Asheville, North Carolina, take Interstate 26 East to US 74 East, follow US 74 East towards the town of Shelby, turn left onto SR 1162 going towards the town of Lattimore, turn left at stop sign onto SR 1326-1325, follow to SR 1337 (Zion Church Road), turn right, drive to the Kenneth Whittaker residence (3225) on the right. Park here and pay for collecting. If no one is home, leave money under the rock on the porch.

WHAT TO LOOK FOR: After you pay the owner, go through the gate to the left of the house and walk down into the woods. You will cross a creek on a foot log bridge, continue up the hill on the other side about 200 yards, you will see the collecting area. Here you need to dig down about 3–5 feet in the soft dirt.

You will find nice bottle glass green muscovite single crystals and clusters, you may also find blue anatase, red rutile needle crystals and smoky and phantom quartz crystals.

FEE: There is a $3.00 per-person per-day fee to collect here.

SAFETY: This is a safe place to collect.

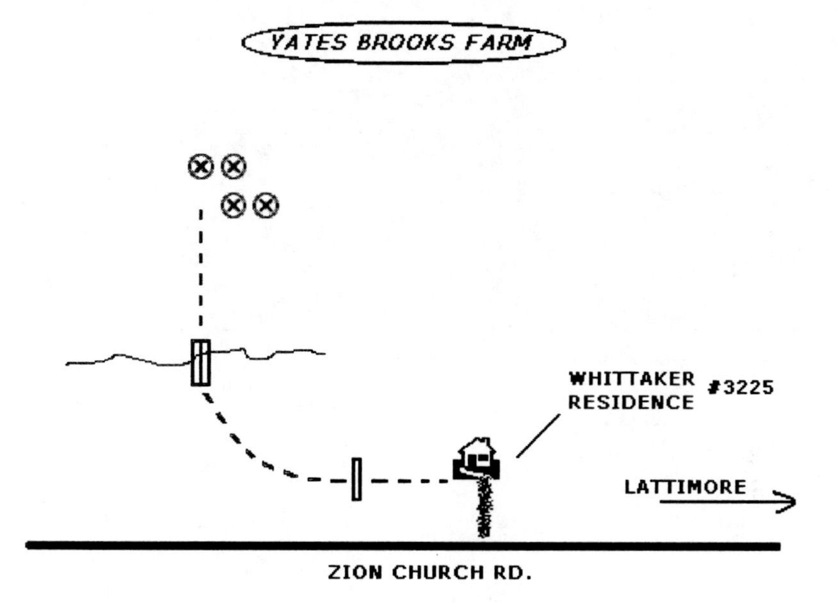

Cluster of muscovite crystals from the Yates Brooks Farm

SITE 50: PORTER LIMONITE

LOCATION: Stanly County, North Carolina.

BEST SEASON: Any, weather permitting.

PROPERTY OWNER: Private (Bowers family).

MATERIAL TO COLLECT: Limonite pseudomorph cubes after pyrite.

TOOLS: Shovel, rock pick, 1/2" sifting screen.

VEHICLE: Any.

DIRECTIONS: From Asheville, North Carolina take Interstate 26 East into South Carolina, follow to Interstate 85 North, follow I-85 North through Charlotte, North Carolina to the Albemarle exit, take US Highway 52 South through Albemarle to the town of Porter, turn right onto Stanly School Road (SR 1923), follow to Cottonville Road (SR 1918), turn right, cross railroad tracks, look on the right, brick house (Bowers) on mailbox, stop here to pay owner.

WHAT TO LOOK FOR: Follow the road to the right of the owners' property to the woods behind the house approximately 300 yards. There is a cut on top of the hill where the Michelin plant has installed an oil line. You will see prospecting holes throughout the area. You can dig and sift to find limonite cubes ranging in size from 1/8" up to 4". You may also find nice matrix pieces, to find the larger cubes you need to dig down 3-5 feet, it is fairly easy digging. Cubes up to 1" are common and can be picked up from the surface.

FEE: The fee to dig is $10.00 per-person per-day. You pay Mr. Bowers at his residence. If he is not home, you can leave the money in a can on the porch.

SAFETY: This is a safe place to collect.

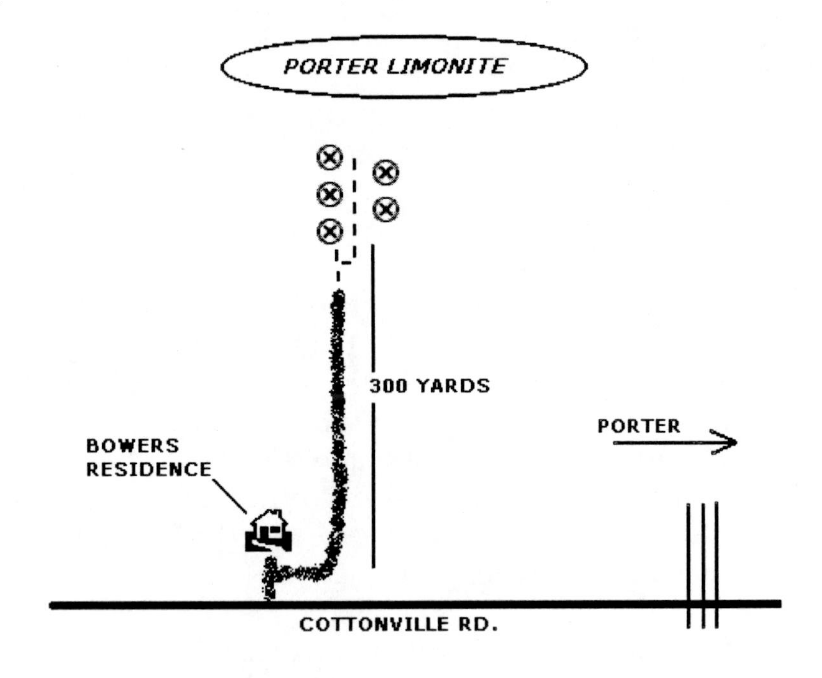

Specimens of limonite from the Porter site

SITE 51: CARTER MINE

LOCATION: Buncombe County, North Carolina.

BEST SEASON: Any, weather permitting.

PROPERTY OWNER: Private (Marilyn Ayers).

MATERIAL TO COLLECT: Pink and white corundum, common opal.

TOOLS: None.

VEHICLE: Any.

DIRECTIONS: From Asheville, North Carolina, take Highway 19-23 North to the Jupiter/Barnardsville/197 exit. Turn right and drive 3.9 miles on NC 197 towards Barnardsville, turn left onto Holcombe Branch Road (SR 2162), drive 1.6 miles, turn right into the Seven Glens subdivision on Seven Glens Drive, follow 0.1 mile to Cove Creek Lane, turn left and drive 0.3 miles to Meadow Creek Drive, turn left and drive 0.1 mile and park out of the road.
GPS Coordinates: 35 48.410 N 082 28.855 W

WHAT TO LOOK FOR: This site was once the location of the Carter Corundum Mine. The land had been in the Carter family since the 1800s, but was recently sold to a developer and turned into a subdivision. The site I have listed here is the location of the mine dumps. Look to your right (east) from where you are parked and you will see a small hill on the other side of a small stream. Cross the stream and walk about 100 feet up the hill to the tree at the top. From the tree search down the hill on the south side to find numerous pieces of pink and white corundum and you may also find some common opal. The lady who owns the property does not mind surface collecting as long as you ask permission first. She lives at the house just south of the hill at No. 80 Meadow Creek Drive.

FEE: There is no fee to collect at this site, make sure you ask permission from the owner before collecting.

SAFETY: This is a safe place to collect.

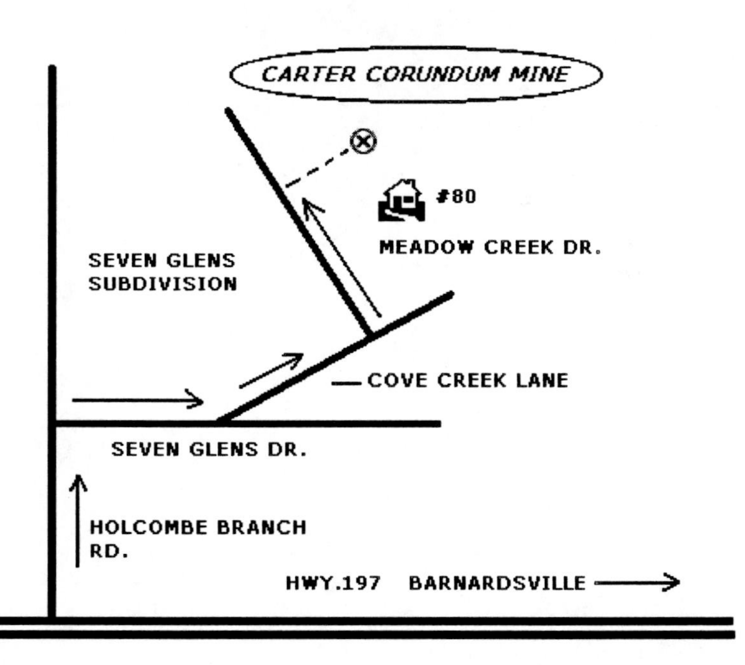

Cabachon cut from pink corundum and blue serpentine from the Carter site

SITE 52: ERWIN CALCITE

LOCATION: Unicoi County, Tennessee.

BEST SEASON: Any, weather permitting.

PROPERTY OWNER: State of Tennessee, Dept. of Transportation.

MATERIAL TO COLLECT: Dogtooth calcite crystals, flint, pyrite.

TOOLS: Rock pick, 3-lb. sledgehammer, rock chisel.

VEHICLE: Any.

DIRECTIONS: From Asheville, North Carolina, take Highway 19-23 North to exit 9, continue straight onto Highway 23 North, drive up the mountain 10.5 miles to the North Carolina/Tennessee border. From the Tennessee state line drive 14.3 miles to exit 15 Jackson Love Hwy./Erwin, go left across the bridge back over the highway, turn onto Highway 23 South going back towards North Carolina, drive 0.8 miles to the first road cut on the right and park out of the road.

WHAT TO LOOK FOR: This is another Tennessee site. It was first discovered by rockhound Steve Penley when the road was being constructed. It is a short drive from the North Carolina border. Look for veins of the reddish brown host rock in the road cut. You will find nice white and clear dogtooth calcite crystals up to 1/4" in length. Search a little and you may also find various colors of flint and some small pyrite crystals.

FEE: There is no fee to collect here.

SAFETY: This is a safe place to collect, stay away from the highway.

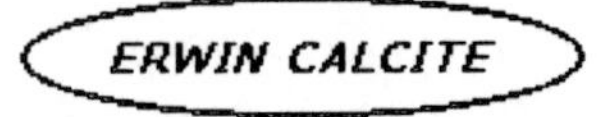

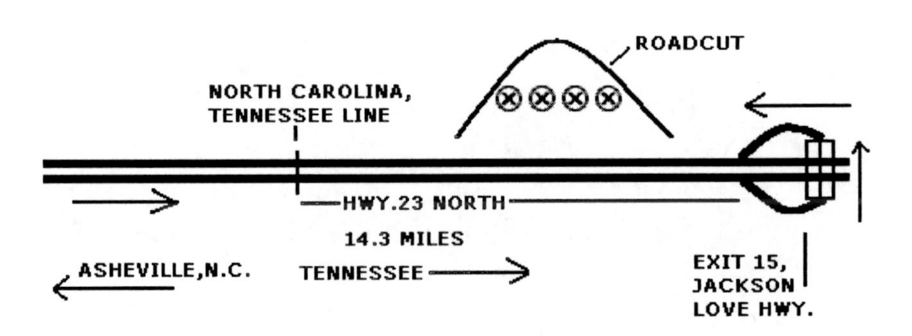

Dogtooth calcite crystals in matrix from the Erwin site

Dig in this road cut to find nice dogtooth calcite specimens

SITE 53: BALSAM MICA MINE

LOCATION: Yancey County, North Carolina.

BEST SEASON: Any, weather permitting.

PROPERTY OWNER: National Forest Service.

MATERIAL TO COLLECT: Beryl, tourmaline (schorl), apatite, garnet, mica, gem feldspar.

TOOLS: 3-lb. sledgehammer, rock chisels, rock pick, shovel, 1/2" sifting screen.

VEHICLE: Any to forest road, four-wheel drive to locked forest gate.

DIRECTIONS: From the Ray Mine site return to Bolens Creek Road, turn left and drive 1.0 mile to a left turn onto a gravel road, this road is just past Stanleys Trl. Road. Follow the gravel road 0.1 miles to the entrance to the forest road. You will need to park here, if you have a four-wheel drive vehicle you can continue 0.1 mile to the locked forest gate and park near the gate.
GPS Coordinates: 35 52.535 N 082 17.012 W (parking area)
GPS Coordinates: 35 52.05 N 082 16.09 W (site)

WHAT TO LOOK FOR: If you like hiking as well as rock hunting this is the site for you. The Balsam Mica Mine is near the top of this mountain at the head of Colberts Creek. You will need to hike to the mine from the parking area at the locked forest gate, only hiking is permitted up the mountain, it is just under 3 miles to the mine and dump piles. I will give you approximate directions to the mine: From the gate it is approximately 1200 yards south to the first road fork, the right fork will lead to the creek, stay to the left going up the mountain. It is another 500 yards and the road will bear left (south east) up the mountain, continue approximately 800 yards. There is a National Forest boundary marker at the top of the hill. The road will turn to the right (east) continue for another 300 yards and you will see a huge tree with four trunks growing from it on your right, this is a good landmark. Continue another 1600 yards to another fork in the road, again the right fork leads to the creek, stay to the left up the mountain for approximately 550 yards. Here you need to watch the road for signs of pegmatite material in the road, when you see it turn right into the woods and walk about 50 yards and you will see the dump piles. Be careful when looking for the mine—it is a steep drop down to the shaft, so pay attention. If you did not find the mine continue to the top of the hill, you will see a grassy clearing in a field on the right side of the road with a campsite and fire pit, if you come to this you have gone too far. Hike back down the mountain about 500 feet and back into the woods and you will locate the mine. This area is not hunted a lot because of the hike to get to it so there is still plenty of material to find. I would break the larger rocks to find beryl or sift the dirt for smaller crystals. While I was at this site I found gem quality feldspar (blue moonstone) and almandine garnet on top of the dump piles.

FEE: There is no fee to collect at this site, forest service rules apply.

SAFETY: This is a remote location in the woods so be aware of wildlife: bears, mountain lions, snakes, etc. Be careful when walking through the woods to the mine, there is no fence around the top and it is a long fall into the pit.

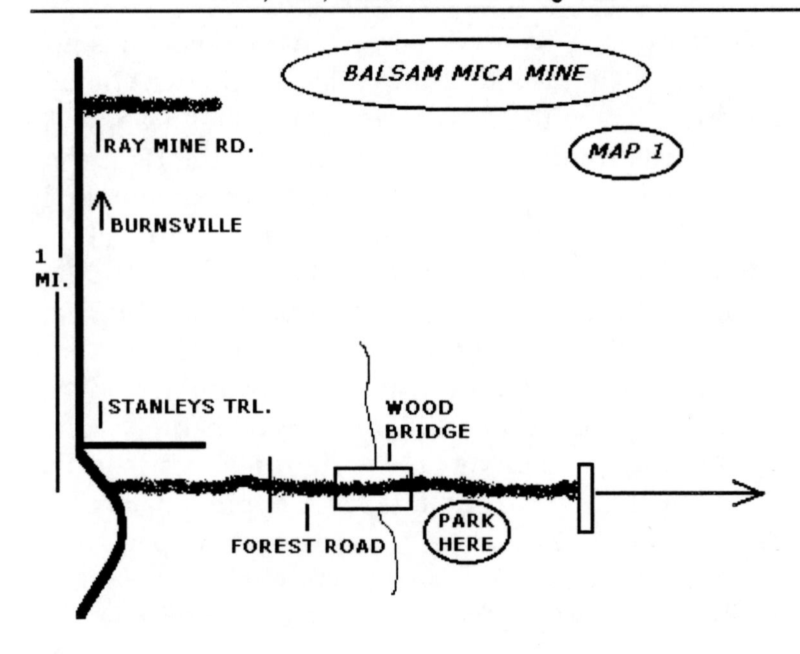
BALSAM MICA MINE
MAP 1
RAY MINE RD.
BURNSVILLE
1 MI.
STANLEYS TRL.
WOOD BRIDGE
FOREST ROAD
PARK HERE

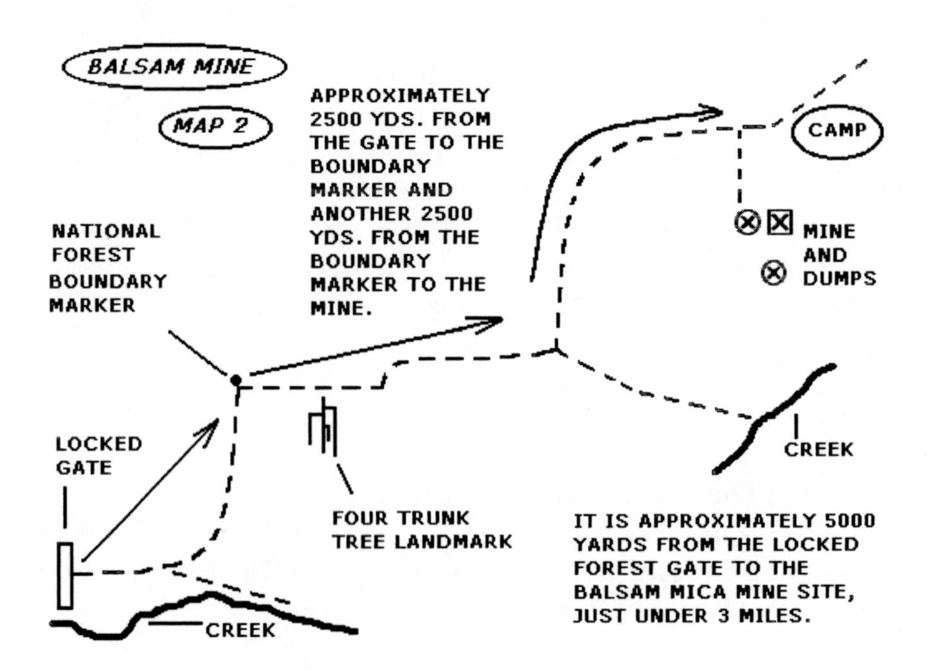
BALSAM MINE
MAP 2
APPROXIMATELY 2500 YDS. FROM THE GATE TO THE BOUNDARY MARKER AND ANOTHER 2500 YDS. FROM THE BOUNDARY MARKER TO THE MINE.
CAMP
NATIONAL FOREST BOUNDARY MARKER
MINE AND DUMPS
LOCKED GATE
CREEK
FOUR TRUNK TREE LANDMARK
IT IS APPROXIMATELY 5000 YARDS FROM THE LOCKED FOREST GATE TO THE BALSAM MICA MINE SITE, JUST UNDER 3 MILES.
CREEK

This tree is a good landmark on the way to the Balsam collecting site

It is a long fall into the water filled pit at the Balsam mine site, so be careful!

Single kyanite crystals and masses like these can be found in the road on the way to the Balsam Mine

FACTS ABOUT NORTH CAROLINA ROCKS, GEMS AND MINERALS

The official state stone is emerald.

The official state rock is granite.

Today, mining is still a half-billion dollar industry in North Carolina.

North Carolina produces the best quality and largest emeralds in North America.

The first gold discovery in the United States was made in North Carolina in 1799.

The North Carolina mountains are over a billion years old, and Mount Mitchell is the highest mountain east of the Mississippi River at 6,684 feet.

To date, 13 diamonds have been found in North Carolina.

During wartime, North Carolina was the only producer of corundum in the country.

Today, over 310 mineral species are known to come from North Carolina.

The largest star sapphire in the world was found in North Carolina.

ABOUT THE AUTHOR:

Richard "Rick" Jacquot was born in Montgomery County, Maryland in 1962, and grew up in Wheaton, Maryland. As a child he and his father, mother, and sister went on numerous camping trips to the mountains of Pennsylvania and West Virginia, this is where the rockhound bug first bit him (actually, back then he was just a rockpup). Being in prime rockhound territory it was not hard even for a ten year old to find many nice specimens (at least they looked nice to him). Eventually he had gathered a decent collection. In 1987, at the age of twenty-five, he moved to the Western North Carolina mountains and found work as a truck driver and later a police officer. Soon he realized that he was in the heart of one of the best rockhound areas in the country and the bug bit him again. Today, fifteen years later, he lives in Leicester, N.C. and has put together a large collection of North Carolina gems and minerals as well as specimens from around the country and the world. He and his son R.J. operate "Jacquot and Son Mining" selling wholesale and retail gem and mineral specimens. They also conduct guided field collecting trips to local mines in the area.

If you are interested in a trip you can contact him at email address: RJacquot@msn.com

The author collecting travertine (cave onyx) at Ladds Mountain in Cartersville, Georgia